SpringerBriefs in Molecular Science

Ultrasound and Sonochemistry

Series editors

Bruno G. Pollet, Faculty of Engineering, Norwegian University of Science and Technology, Trondheim, Norway

Muthupandian Ashokkumar, School of Chemistry, University of Melbourne, Melbourne, VIC, Australia

SpringerBriefs in Molecular Science: Ultrasound and Sonochemistry is a series of concise briefs that present those interested in this broad and multidisciplinary field with the most recent advances in a broad array of topics. Each volume compiles information that has thus far been scattered in many different sources into a single, concise title, making each edition a useful reference for industry professionals, researchers, and graduate students, especially those starting in a new topic of research.

More information about this series at http://www.springer.com/series/15634

About the Series Editors

Bruno G. Pollet is a full Professor of Renewable Energy at the Norwegian University of Science and Technology (NTNU) in Trondheim. He is a Fellow of the *Royal Society of Chemistry* (RSC), an Executive Editor of *Ultrasonics Soncohemistry* and a Board of Directors' member of the *International Association of Hydrogen Energy* (IAHE). He held Visiting Professorships at the University of Ulster, Professor Molkov's HySAFER (UK) and at the University of Yamanashi, Professor Watanabe's labs (Japan). His research covers a wide range of areas in Electrochemical Engineering, Electrochemical Energy Conversion and Sono-electr ochemistry (Power Ultrasound in Electrochemistry) from the development of novel materials, hydrogen and fuel cell to water treatment/disinfection demonstrators & prototypes. He was a full Professor of Energy Materials and Systems at the University of the Western Cape (South Africa) and R&D Director of the National Hydrogen South Africa (HySA) Systems Competence Centre. He was also a Research Fellow and Lecturer in Chemical Engineering at The University of Birmingham (UK) as well as a co-founder and an Associate Director of The University of Birmingham Centre for Hydrogen and Fuel Cell Research. He has worked for Johnson Matthey Fuel Cells Ltd (UK) and other various industries worldwide as Technical Account Manager, Project Manager, Research Manager, R&D Director, Head of R&D and Chief Technology Officer. He was awarded a Diploma in Chemistry and Material Sciences from the Université Joseph Fourier (Grenoble, France), a B.Sc. (Hons) in Applied Chemistry from Coventry University (UK) and an M.Sc. in Analytical Chemistry from The University of Aberdeen (UK). He also gained his Ph.D. in Physical Chemistry in the field of Electrochemistry and Sonochemistry under the supervision of Profs. J. Phil Lorimer & Tim J. Mason at the Sonochemistry Centre of Excellence, Coventry University (UK). He undertook his PostDoc in Electrocatalysis at the Liverpool University Electrochemistry group led by Prof. David J. Schiffrin. Bruno has published many scientific publications, articles, book chapters and books in the field of Sonoelectrochemistry, Fuel Cells, Electrocatalysis and Electrochemical Engineering. Bruno is member of editorial board journals (*International Journal of Hydrogen Energy/Electrocatalysis/Ultrasonics Sonochemistry/Renewables-Wind, Water and Solar/Electrochem*). He is also fluent in English, French and Spanish. *Current Editorships: Hydrogen Energy and Fuel Cells Primers Series (AP, Elsevier) and Ultrasound and Sonochemistry (Springer).*

Prof. Muthupandian Ashokkumar (Ashok) is a Physical Chemist who specializes in Sonochemistry, teaches undergraduate and postgraduate Chemistry and is a senior academic staff member of the School of Chemistry, University of Melbourne. Ashok is a renowned sonochemist, with more than 20 years of experience in this field, and has developed a number of novel techniques to characterize acoustic cavitation bubbles and has made major contributions of applied sonochemistry to the Materials, Food and Dairy industry. His research team has developed a novel ultrasonic processing technology for improving the functional properties of dairy ingredients. Recent research also involves the ultrasonic synthesis of functional nano- and biomaterials that can be used in energy production, environmental remediation and diagnostic and therapeutic medicine. He is the Deputy Director of an Australian Research Council Funded Industry Transformation Research Hub (ITRH; http://foodvaluechain.unimelb.edu.au/#research; Industry Partner: Mondelez International) and leading the Encapsulation project (http://foodvaluechain.unimelb.edu.au/research/ultrasonic-encapsulation). He has received about $ 15 million research grants to support his research work that includes several industry projects. He is the Editor-in-Chief of *Ultrasonics Sonochemistry*, an international journal devoted to sonochemistry research with a Journal Impact Factor of 4.3). He has edited/co-edited several books and special issues for journals; published ~ 360 refereed papers (H-Index: 49) in high impact international journals and books; and delivered over 150 invited/keynote/plenary lectures at international conferences and academic institutions. Ashok has successfully organised 10 national/international scientific conferences/workshops and managed a number of national and international competitive research grants. He has served on a number of University of Melbourne management committees and scientific advisory boards of external scientific organizations. Ashok is the recipient of several prizes, awards and fellowships, including the Grimwade Prize in Industrial Chemistry. He is a Fellow of the RACI since 2007.

Rachel Pflieger · Sergey I. Nikitenko ·
Carlos Cairós · Robert Mettin

Characterization
of Cavitation Bubbles
and Sonoluminescence

 Springer

Rachel Pflieger
Marcoule Institute for Separation Chemistry
ICSM UMR5257, CEA, CNRS
University of Montpellier, ENSCM
Bagnols-sur-Cèze Cedex, France

Sergey I. Nikitenko
Marcoule Institute for Separation Chemistry
ICSM UMR5257, CEA, CNRS
University of Montpellier, ENSCM
Bagnols-sur-Cèze Cedex, France

Carlos Cairós
Department of Analytical Chemistry
University of La Laguna
La Laguna, Tenerife, Spain

Robert Mettin
Third Institute of Physics
Georg-August-University Göttingen
Göttingen, Germany

ISSN 2191-5407   ISSN 2191-5415   (electronic)
SpringerBriefs in Molecular Science
ISSN 2511-123X   ISSN 2511-1248   (electronic)
Ultrasound and Sonochemistry
ISBN 978-3-030-11716-0   ISBN 978-3-030-11717-7   (eBook)
https://doi.org/10.1007/978-3-030-11717-7

Library of Congress Control Number: 2018967437

This Springer imprint is published by the registered company Springer Nature Switzerland AG
The registered company address is: Gewerbestrasse 11, 6330 Cham, Switzerland

Preface

Acoustic cavitation is a peculiar phenomenon that occurs in liquids irradiated with strong acoustic fields: Bubbles appear that are subsequently driven by the field to undergo strong volume pulsations. Since the conditions in and near the bubbles can become really extreme during their implosion, a variety of surprising and intriguing phenomena can be observed in cavitating liquids. And right after discovery, people tried to use such effects and employ cavitation as a tool—usually with ultrasound. Possibly, the most prevalent and best-known application is ultrasonic cleaning, but another branch has gained significant attention and importance over the last decades: cavitation chemistry, better known as *sonochemistry*. Although acceptance and transfer of sonochemical methods to industrial processing might have started more reserved than it was the case in ultrasonic cleaning, nowadays, cavitation chemistry is a strong and recognized chemistry branch, and its exotic flavor has ceased. Nevertheless, cavitation, bubble dynamics, and the processes in collapsing bubbles are complex. And in particular, the interdisciplinary character of these subjects requires an open mind to learn and to train—even for the specialists in one of the connected fields, as there are chemistry, chemical engineering, material science, acoustics, hydrodynamics, and more. There are in the meantime excellent monographs, article collections, and review papers available for a thorough study of virtually every relevant topic. However, possibly the chance for a rapid overview with a final up-to-date picture of sonochemistry and the related topics is missing. Therefore, the aim of this little book is not to supply a complete reference nor to replace established textbooks, but to give the reader a concise and modern introduction at hand for a qualified overview. The book is part of the series *Springer Briefs in Molecular Science: Ultrasound and Sonochemistry*, and the other titles can perfectly complement our text.

The book focuses on the characterization of cavitation bubbles with respect to multibubble systems, either through direct observations of the bubbles or through measurements of sonoluminescence spectra and of sonochemical activity. Chapter 1 gives a short background on nucleation and dynamics of individual bubbles in a sound field, proceeds with bubble instabilities and interactions, and finally examines bubble ensembles. Chapter 2 introduces to sonoluminescence, its interpretation, and spectral analysis of the hot plasma core of collapsed bubbles. Methods for derivation of pressures and plasma analysis are presented, and pathways for the chemistry behind the emissions are discussed, along with bubble sizing methods based on sonoluminescence. Chapter 3 discusses sonochemistry in terms of dosimetry, chemical reactions, and dependences on parameters like ultrasonic frequency and important chemical additives.

We hope the reader finds the text useful and profits from the topics selected. The field is huge, and we are aware that many important aspects remain only touched or even untold. Thus, we encourage the reader to follow the literature hints and references of his or her interest to deepen and complete the knowledge on ultrasound, cavitation, sonoluminescence, and sonochemistry.

The authors would like to thank Bruno Pollet and Muthupandian Ashokkumar for encouraging this contribution to the Springer Briefs series.

R. P. and S. N. thank Tony Chave, Matthieu Virot and all actual and former students and postdocs of the sonochemistry group at ICSM for their valuable contributions in the studies performed at ICSM. Obviously, many thanks go to people from whom we learned a lot during enriching collaborations: Muthupandian Ashokkumar (University of Melbourne), Robert Mettin, Carlos Cairós, and Thomas Kurz (Georg-August-University Göttingen), Micheline Draye (Savoie Mont Blanc University), Thierry Belmonte (Lorraine University) and their teams.

R. M. and C. C. thank all actual and former members of the cavitation and bubble dynamics group at Drittes Physikalisches Institut, Georg-August-University Göttingen. Without their contributions over decades, the map of acoustic cavitation would still contain many more white spots, and it is a pleasure to report on results from the group. Particular thanks go to Werner Lauterborn, Thomas Kurz, Andrea Thiemann, Till Nowak, Fabian Reuter, Philipp Frommhold, Reinhard Geisler, Max Koch, and Christiane Lechner, to name just a few of the many important people. Especially, we like to thank the Christian Doppler Forschungsgemeinschaft and Lam Research AG Villach (Austria) for continuous and generous support in the framework of the Christian Doppler Laboratory for Cavitation and Micro-Erosion and particularly Frank Holsteyns, Alexander Lippert, and Harald Okorn-Schmidt.

Finally, the authors would like to thank all staff at Springer for their qualified support and patience.

Bagnols-sur-Cèze, France   Rachel Pflieger
Bagnols-sur-Cèze, France   Sergey I. Nikitenko
La Laguna, Spain   Carlos Cairós
Göttingen, Germany   Robert Mettin

Contents

Chapter 1
Bubble Dynamics

1.1 Introduction

Bubble dynamics and cavitation have been recognized as a relevant topic of physics and engineering for more than 100 years. Starting with erosion problems at ship propellers end of the nineteenth century [1, 2], experimental and theoretical research went on to intense ultrasound fields in liquids after World War I [3]. However, the phenomena are intrinsically difficult to investigate since the involved spatial scales span many orders of magnitude, the timescales are partly extremely fast, and the behavior of bubbles includes important nonlinearities. Thus, it is not surprising that the development of the subject was fostered significantly by advances in technical equipment, namely high-speed imaging devices and modern computers. In the meantime, there exist several excellent review articles and monographs on cavitation. Starting with the earlier work by Flynn [4] and Rozenberg [5] from the 1960s and 1970s, they proceed within the 1980s with Neppiras [6] and Young [7], to be followed by Leighton [8] and Brennen [9] in the 1990s. Till today, an amazing progression of research activity in the whole field can be found, boosted by the strong interest in sonoluminescence (see the overview by Young [10]) and the perspective of non-classical chemistry (see for instance Mason's and Lorimer's books on sonochemistry [11, 12]). Due to the growth and spreading of the topic, many review articles and specialized overviews have appeared meanwhile, and an adequate and complete reference is difficult, if not impossible. From the perspective of the quite active group at Göttingen University, several recent reviews and collections of previous work should be mentioned; see [13, 14–18]. The following chapter does not intend to replace a deeper study of the indicated references on cavitation, but wants to give a rapid and useful overview of characteristics of acoustic cavitation bubbles. This serves as a certain basis for the subsequent chapters on sonoluminescence and sonochemistry in multibubble systems, but also as a general briefing, facilitating further studies.

© The Author(s), under exclusive licence to Springer Nature Switzerland AG 2019 1
R. Pflieger et al., *Characterization of Cavitation Bubbles and Sonoluminescence*,
Ultrasound and Sonochemistry, https://doi.org/10.1007/978-3-030-11717-7_1

1.2   The Bubble Collapse

The key phenomenon that leads to the most spectacular effects of cavitation is the rapid and strong compression of gas phase in a bubble, also termed a *bubble collapse*. This leads to dramatic pressure and temperature peaks that in turn cause acoustic shock waves and erosion of hardest materials outside the bubble and chemical reactions and light emission in the interior. Rapid compression and heating of gas or plasma is known in various contexts and on quite different scales, e.g., supernova explosions and interstellar shock waves, atmospheric re-entrance and supersonic flight, combustion engines or inertial confinement fusion. In cavitation, the peculiar agent of compression is rather simple and usually small: a *bubble*, i.e., a "void" in a liquid.

1.3   Bubble Size

Bubbles are not really empty, but usually contain vapor of the host liquid and non-condensable gas. While the ambient liquid is usually to a good approximation incompressible, the gaseous content of the bubble can be well expanded or squeezed, which leads to the potential of substantial changes in bubble volume. It is convenient to start considerations of dynamics with an idealized, spherical bubble embedded in a three-dimensional infinite domain of liquid. The bubble radius is denoted by $R(t)$ and time by t. The liquid pressure far away from the bubble is named $p_\infty(t)$. It contains the hydrostatic pressure p_0 and any acoustic pressure caused by an additional sound field. The external pressure variations govern the oscillating bubble behavior, but let us first have a look onto the static case. For constant $p_\infty(t) = p_0$, there exists an equilibrium; i.e., the bubble is at rest at its *equilibrium radius* R_0, also called *rest radius*. Then, the outside liquid pressure is compensated by the interior gas pressure p_b, composed of the pressure of non-condensable gas p_g (e.g., air) plus the vapor pressure p_v that also counteracts from the bubble inside. To consider the surface tension σ between liquid and gas phase, an additional outside pressure $p_s(R) = 2\sigma/R$, the *Laplace pressure*, comes into play [9]. The equilibrium radius including these effects results from the pressure balance of inside and outside pressures: $p_g(R_0) + p_v = p_s(R_0) + p_0$. We denote $p_g(R_0) = p_{g,0}$ and assume that the interior bubble pressure is solely determined by the actual bubble radius.[1] This follows, for instance, for a given amount of homogeneous non-condensable ideal gas with the equation of state $pV = NR_gT$, with the gas constant R_g, the temperature T, and the bubble volume $V = 4\pi R^3/3$. At the equilibrium radius and an ambient temperature T_0, one obtains

---

[1]More realistic descriptions of bubbles might consider non-equilibrium conditions like heat conduction, inhomogeneous bubble interior, or dynamics of evaporation/condensation of liquid/vapor at the bubble wall.

$p_{g,0}4\pi R_0^3/3 = NR_gT_0$. Then, the number of moles N of the gas is related to the equilibrium radius via the equation $N = 4\pi[R_0^3(p_0 - p_v) + 2\sigma R_0^2]/(3R_gT_0)$. Note here that due to the surface tension, the inside bubble pressure is higher than the ambient pressure p_0, which can cause diffusion of gas out of the bubble.

Since in general cases the bubble can undergo large variations of its volume in time, it is important to specify what exactly is meant whenever one is talking about the bubble "size". The equilibrium radius R_0 from the static case is usually (and in the following) employed to measure the "equilibrium size" of a bubble (equivalently one could, for instance, refer to the static equilibrium volume or to the amount of non-condensable gas molecules in the bubble). Other important bubble size measures are the *maximum radius* R_{max} and the *minimum bubble radius* R_{min} during a volume oscillation. These can be quite distinct from the equilibrium value, and determine the expanded and compressed state.

1.4 Nuclei and Nucleation

Where do the bubbles come from? The formation of a bubble can occur under various conditions. Physically speaking, distinct cases are *boiling*, i.e., a local energy deposition and evaporation of a certain liquid volume under ambient pressure conditions, and *cavitation*, where the liquid evaporates due to a tension (negative pressure) in the liquid under ambient temperature. If the tensile stress is caused by bulk liquid flows, one speaks of *hydrodynamic cavitation*, and if it is generated by a sound wave, we study *acoustic cavitation*. The fundamental phenomena of the two types of cavitation are mostly the same, but it is the latter mechanism that we will focus on.

It is important to note that as long as one is dealing with so-called *real liquids*, in particular water, there typically exist stabilized (sub)micron-sized entities of non-condensable gas (e.g. air), even under silent conditions [19–21]. Such small pre-existing bubbles are termed *nuclei*, and their abundance and size distribution depend on parameters like dissolved gas content or temperature, but also on the history of the liquid sample.[2] As an example, Fig. 1.1a shows measured nuclei statistics of untreated, of degassed, and of filtered tap water.

The first occurrence of cavitation bubbles in a liquid under tensile excitation is called *nucleation*, and any parameter threshold that is passed to cause nucleation is termed *cavitation threshold*. However, in all cases where no extremely clean and degassed liquid is used and no special pretreatment has been undertaken, cavitation bubbles occur in a process that might rather be termed "activation" than "nucleation": Some of the previously passive nuclei, stabilized in the bulk liquid, at

---

[2]Free submicron bubbles should dissolve quite rapidly because of surface tension, as suggested above. However, bubbles might be stabilized in crevices of solid particles [8] or be stabilized statically or dynamically when covered partly with hydrophobic material; see [8, 22].

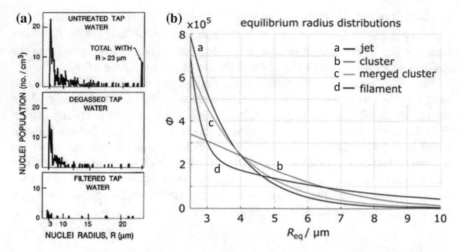

Fig. 1.1 a Histograms of nuclei populations in treated and untreated tap water (reproduced from [23] and [9], with kind permission). Toward smaller sizes, the number density of nuclei typically increases. However, also, the measurements get more difficult, and thus, the statistics shown are cut off and do not include nuclei radii below about 1 μm. Such smaller nuclei should as well exist abundantly. **b** Equilibrium bubble radius distribution obtained from a cavitating system under 25 kHz sonication in tap water $(R_{eq} = R_0)$. Letters (colors online) indicate different cavitation bubble structures. Reproduced from [24], © Elsevier 2018

contaminants or container walls, are turned into expanding (and subsequently collapsing) cavitation bubbles by the tensile stress (caused by the acoustic wave in acoustic cavitation). In this sense, we deal usually with *heterogeneous nucleation*. This explains the experience that cavitation thresholds in terms of tensile strength or acoustic pressure amplitudes are typically much lower than their values expected for a tensile rupture of the pure liquids (which would be "real" cavitation in the sense of *homogeneous nucleation*; compare [9]). The reason is that for expansion of a pre-existing void of radius R, a tension of the order of the Laplace pressure $2\sigma/R$ is needed, and "pre-existing voids" in a pure liquid, caused by random fluctuations, would have radii R not much larger than typical intermolecular distances—leading to extreme tension values. The presence of nuclei in non-treated real liquids (like tap or sea water, Fig. 1.1a) reduces the needed tension to the order of $2\sigma/(1\mu m)$, which in the case of water yields about 150 kPa since $\sigma \approx 0.075$ N/m. In effect, cavitation thresholds and also the numbers of nucleated cavitation bubbles in real liquids will depend on liquid impurities, dissolved gas, or even the vessel walls. Such factors, which are not always easy to control, give rise to some degree of sensitivity and variability of cavitating systems, experienced as a certain amount of randomness. In addition, a *cavitating* system will change the nuclei distribution *by itself* due to gas diffusion out of or into the oscillating bubbles and merging or splitting bubble events. Assessment of equilibrium bubble radii under sonication conditions is, however, not an easy task, since the visible bubble sizes are permanently changing. Results from a recent approach based on the statistics of momentary bubble sizes and a single-bubble model [24] are shown in Fig. 1.1b. Comparison of these distributions of cavitation bubble equilibrium radii with the distributions of still nuclei reveals that the decaying shape and the order of absolute magnitude (several µm) essentially remain the same. Differences occur in the details; e.g., less large bubbles are found in the strong sound field since they split and break up.

1.5   Bubble Oscillations, Nonlinearity, and the Rayleigh Collapse

If we imagine a gentle variation of the bubble radius from its equilibrium size, it gets clear that bubbles can *oscillate* around R_0: Expansion leads to a decrease of interior pressure and thus to the tendency to restore the original bubble size. Vice versa, a compression will increase the gas pressure and lead to a rebound. The bubble volume finally forms an oscillatory system, and its own resonance frequency can be derived as

$$f_{res} = \frac{1}{2\pi R}\left(\frac{3\gamma p_0}{\rho}\right)^{1/2}, \quad R_{res} = \frac{1}{2\pi f}\left(\frac{3\gamma p_0}{\rho}\right)^{1/2} \tag{1.1}$$

This frequency f_{res} is the linear resonance of a bubble, also called the Minnaert frequency [9, 25]. The value depends on the ambient pressure p_0, the liquid density ρ, and the polytropic exponent γ of the ideal gas. On the right side, we have inverted the formula to arrive at an expression for a linear resonant radius R_{res}. This is of importance if a fixed driving frequency acts on a population of bubbles, a case usually met in technical applications of ultrasound and in particular in sonochemistry.

Resonant bubble oscillations are, for instance, responsible for audible sound of running or splashing water [25]. It turns out, however, that this bubble oscillator is *nonlinear*: While one could imagine an arbitrary large expansion of a bubble, its compression below zero radius is unphysical. This is one reason for the restoring force being asymmetric with respect to expansion and compression, which finally leads to anharmonic oscillations [26]—including the implosive "collapse" type of motion. In fact, the most extreme case of a bubble collapse would happen if the spherical cavity was *empty*; i.e., no gas (or vapor) pressure would ever counteract the inflowing liquid (and thus no rebound would occur). Historically, Rayleigh [2] was the first to investigate the spherical implosion into empty space, which is why it is called *Rayleigh collapse*. The speed of the spherically inrushing liquid front as well as the pressure in the liquid (assumed to be incompressible) can be calculated analytically for this case. It is found that an infinite speed and infinite pressure in the liquid at the bubble would be approached at the cavity closing moment [2, 7]. This result serves to illustrate the enormous energy focusing of a bubble collapse by the three-dimensional influx and in particular the erosive potential by the pressure peak. However, for a description of the internal conditions of realistic bubbles during their compression phase, consideration of the gas is additionally required (and Rayleigh did so as well in his paper [2]). Then, the collapse is cushioned to finite—but potentially still quite high—wall velocities, and finally, the inward liquid motion is stopped and reversed. The gas is rapidly heated up during the implosion, and high pressure and temperature peaks occur at the minimum bubble size, which facilitates chemical reactions and light emission from the gas—*sono-chemistry* and *sonoluminescence*. In Fig. 1.2a and b, a Rayleigh collapse of an empty bubble is illustrated, showing the bubble radius and the diverging bubble wall speed. If the same bubble contains a noble gas following an adiabatic ideal gas equation of state, the results for radius–time and bubble wall velocity versus time look like Fig. 1.2c, d if the amount of gas corresponds to an equilibrium radius of $R_0 = 5$ μm. Additionally, gas temperature and gas pressure are plotted versus time in Fig. 1.2e, f. From the adiabatic law and neglecting vapor and surface tension, one obtains $T = T_0(R_0/R)^{3(\kappa-1)}$ and $p = p_0(R_0/R)^{3\kappa}$ with the adiabatic exponent κ (see below) and the reference values $T_0 = 293$ K and $p_0 = 10^5$ Pa. The peaks at

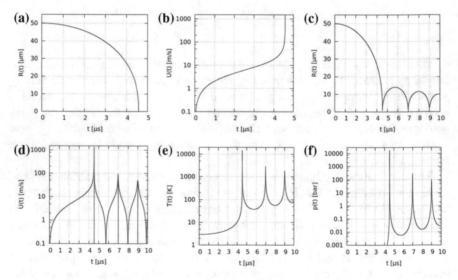

Fig. 1.2 Bubble dynamics in water. The radius–time curve of a Rayleigh collapse of an empty bubble is shown in (**a**). Initially, the radius is 50 μm with a resting bubble wall; the ambient pressure is 100 kPa. **Plot b** shows the modulus of the bubble wall velocity on a logarithmic scale versus time for the empty bubble. The same collapse conditions for an argon-filled bubble with adiabatic heating are shown in (**c**). In the collapse, the gas is compressed and heated, and then, it expands again. Several decaying rebound oscillations occur with less violent collapses. The bubble wall velocity (modulus) is shown in (**d**); note the inversion of direction at the bubble radius minima and maxima, visible as vertical lines in the logarithmic scale. The temperature and the pressure in the bubble, according to ideal adiabatic compression, are shown in plots (**e**) and (**f**)

15,000 K and almost 20 kbar are rather high, but such extreme values occur only during a short moment of time—here several nanoseconds. More realistic views of the bubble would consider in particular heat conduction of the hot gas toward the liquid, and thus, the given values are upper bounds.

1.6 The Rayleigh–Plesset Equation

A quite good model of the dynamics of a *spherical* bubble is obtained by the Rayleigh–Plesset (RP) equation. This ordinary differential equation of second order describes the evolution of the bubble radius $R(t)$ over time when descriptions of the internal bubble pressure $p_b(R)$ and the pressure far from the bubble $p_\infty(t)$ are given [8]:

$$R\ddot{R} + \frac{3}{2}\dot{R}^2 + \frac{4\mu\dot{R}}{\rho R} + \frac{2\sigma}{\rho R} = \frac{1}{\rho}(p_b(R) - p_\infty(t)) \tag{1.2}$$

Effects of surface tension σ and dynamic viscosity μ are included, ρ is the density of the liquid, and the dots denote differentiation with respect to time as usual. Following our choice above of an ideal gas in the bubble, we derive the internal gas pressure as

$$p_b = (p_0 + 2\sigma/R_0 - p_v)(R_0/R)^{3\gamma} + p_v \tag{1.3}$$

with vapor pressure p_v and polytropic exponent γ. A value $\gamma = 1$ represents isothermal compression, while $\gamma = \kappa = c_p/c_v$ corresponds to adiabatic heating of the gas with the adiabatic exponent κ, derived from the ratio of the specific heats at constant pressure and constant volume, c_p and c_v, respectively. For an ideal gas, this ratio can be expressed by the number of available degrees of freedom f_g of the gas molecules: $\kappa = c_p/c_v = (f_g + 2)/f_g$. For noble gases, one obtains $\kappa = 5/3 \approx 1.67$ which results in most effective heating, and this has been used in Fig. 1.2. For air, containing mainly two-atomic gases, the value is close to $\kappa = 7/5 = 1.40$. In simple models that take heat conduction into account, γ can be approximately tuned to an effective value lying between 1 and κ, depending on the momentary bubble wall speed [27]. The term $p_\infty(t)$ comprises the static pressure p_0 and the acoustic pressure $p_{ac}(t)$ that is usually assumed to be harmonic with frequency f and amplitude p_a: $p_\infty(t) = p_0 + p_{ac}(t) = p_0 + p_a \sin(2\pi f t)$.

From the RP equation, one obtains analytically via linearization the Minnaert frequency f_{res} from (1.1) and the small amplitude bubble response in form of linear sinusoidal oscillations around R_0 in dependence on p_a and f [7, 8]. However, when bubble dynamics gets more involved than small harmonic oscillations or the Rayleigh collapse, analytic results are difficult to obtain and become involved [28], and investigations strongly rely on numerical solutions of the equations [26].

Figure 1.3a exemplifies how the sinusoidal bubble wall motion develops into the collapse type for increased driving amplitude. The accompanying part in Fig. 1.3b depicts the bubble displacement if the incident sound field is a traveling wave, which is discussed in detail below in Sect. 1.20.

To reach a more accurate description of strong collapses and the bubble interior, extensions of the standard RP model exist (such models are often termed "Rayleigh–Plesset-like"). For instance, liquid compressibility is taken into account by employing terms with $M = \dot{R}/c$, M being the Mach number of the bubble wall with respect to the sound velocity in the liquid c [29]. Furthermore, the simple laws of the gas in the bubble (e.g., adiabatic or isothermal ideal gas) can be replaced by more sophisticated choices with excluded volume (e.g., van der Waals gas) and heat

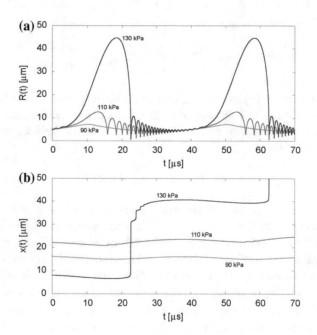

Fig. 1.3 a Radius versus time for a bubble of 5 μm rest radius and sinusoidally driven at 25 kHz. The radius history is shown for increasing driving pressure amplitudes. Up to 90 kPa, the oscillations remain nearly sinusoidally, as in a linear approximation. At 110 kPa, first collapse oscillations occur with characteristic peak form around the minima of the radius. At 130 kPa, after an extended expansion phase to about tenfold the rest radius, a main collapse happens at 22.5 μs. Extreme temperature and pressure conditions occur in this driven collapse, similar to those in the free collapse shown in Fig. 1.2c–f. The sharp qualitative transition in the collapse behavior for only small change in pressure amplitude is characteristic for the nonlinearity of oscillating bubbles and connected with the so-called Blake threshold (see Sect. 1.8). **b** Spatial translation of the bubble from **a** when driven by a plane traveling wave in positive x-direction. At lower excitation (90 and 110 kPa), the bubble is moving forth and back, and only a small mean displacement in direction of wave propagation occurs. At 130 kPa driving, the bubble undergoes a significant and fast jump in forward direction during the main collapse around 22.5 μs, and the net displacement is of the order of the maximum bubble radius before collapse. This is further discusses in Sect. 1.20

conduction models [27]. To include liquid evaporation and condensation as well as chemical reactions, additional equations can be coupled to the radial dynamics [30, 31]. The equations used for Figs. 1.2 and 1.3 and also for the following results follow actually the Keller–Miksis model, an RP variant including liquid compressibility [16, 26, 32]:

$$R\ddot{R}(1-M) + \frac{3}{2}\dot{R}^2\left(1 - \frac{M}{3}\right) = \frac{1+M}{\rho}p_l + \frac{R}{\rho c}\frac{\mathrm{d}p_l}{\mathrm{d}t} \qquad (1.4)$$

$$p_l = \left(p_0 + \frac{2\sigma}{R_0}\right)\left(\frac{R_0}{R}\right)^{3\gamma} - \frac{2\sigma}{R} - \frac{4\mu\dot{R}}{R} - p_0 - p_{\mathrm{ac}}(t) \qquad (1.5)$$

Results presented are based on the numerical solution of this model for water under standard conditions (air pressure and room temperature), either adiabatic $\gamma = \kappa$ or isothermal $\gamma = 1$.

1.7   Acoustic Cavitation: Bubble Types

Let us assume the above introduced pre-existing "nuclei" bubbles as spherical non-condensable gas bubbles of very small radii R_{eq} (although the geometry might be more complicated if they are attached to a microscopic crevice of a dust particle or a wall), and let them be exposed to an external pressure variation. To become "cavitation bubbles" that show relevant effects, some nuclei should expand significantly, and indeed, the characterization of cavitation bubbles by their collapse dynamics is common and convenient (although not necessarily unique in the literature). For instance, the expansion ratio $\alpha = R_{\mathrm{max}}/R_0$ or the compression ratio $\beta = R_0/R_{\mathrm{min}}$ are reasonable dynamical measures for bubble activity in the sense of cavitation effects. From a physical point of view, the bubble wall acceleration is dominated by the inertia of the liquid for $\alpha > 2 \ldots 3$ [6], which leads to a pronounced collapse. Therefore, the term "inertial cavitation" is used frequently to characterize such bubbles with stronger implosion. In contrast, the weaker oscillating bubbles are dominated by the internal gas pressure and are sometimes termed "stable cavitation" ($\alpha < 2 \ldots 3$).[3] With respect to the occurrence of *specific cavitation effects*, the threshold value of α might have to be somehow larger than 3, depending on the effect under question. For instance, in the context of single-bubble sonochemistry, a reaction threshold value of about 4 has been observed [33], and for sonoluminescence in xenon sparged sulfuric acid, α values beyond 6 have been

---

[3]The term "stable" for gas dominated bubble dynamics is somehow unfortunate since less strong collapsing bubbles can nevertheless exhibit instabilities (e.g., develop non-spherical shapes and splitting), while inertial cavitation bubbles can well oscillate in stable regimes. The older notion of "transient" cavitation for inertial cavitation is misleading in the same sense.

found to be consistent with observations of light emission [34]. However, since the dynamical quantities like α or β do not fully determine temperature and pressure peaks during realistic bubble collapses, they should be seen rather as rough indicators than as exact threshold measures.

1.8 The Blake Threshold

The response of a bubble to the external pressure signal depends crucially on its equilibrium radius R_0. To turn a nucleus into a transient cavitation bubble, its surface tension pressure has to be counteracted by the tensile part of the acoustic pressure p_{ac} that together with the static pressure p_0 forms the pressure p_∞ far from the bubble. The application of a quasi-static tension $p_{ac} = -p^* < 0$ will lead to a shift of the bubble size, and the condition for the new equilibrium bubble radius R_0^* reads $p_b(R_0^*) + p_v = p_s(R_0^*) + p_0 - p^*$. Closer analysis shows that from a critical value of tension $-p^* = -p_{cr}^*$ on, a bubble of given equilibrium radius R_0 *cannot* find a new stable equilibrium R_0^* anymore and will expand without bound![4] This phenomenon was described by Blake [35] and Neppiras and Noltingk [36] and is termed the *Blake threshold*. Neglecting vapor pressure, the relation of critical tensile pressure p_{Blake} and bubble rest radius R_0 reads [28, 37]

$$p_{Blake} = p_{cr}^* = p_0 \left[1 + \left(\frac{4}{27} \frac{\alpha_s^3}{(1+\alpha_s)} \right)^{1/2} \right], \quad \alpha_S = \frac{2\sigma}{p_0 R_0} \quad (1.6)$$

Stating this result slightly differently: For a given quasi-static tension $-p^*$, there exists a critical nucleus or bubble equilibrium size $R_{0,cr}$ beyond which an "infinite" expansion occurs. This radius is found by solving (1.6) for R_0 and placing p^* for p_{cr}^*. The Blake threshold phenomenon is encountered as well for the time-varying pressure fields $p_{ac}(t)$ in acoustic cavitation, e.g., the sinusoidal pressure of a sound wave $p_a \sin(\omega t)$ introduced above, here with the angular frequency $\omega = 2\pi f$. Then, the critical tension $-p_{cr}^*$ will occur in the negative-going phase if the driving pressure amplitude surpasses the critical value $p_{a,cr} = p_{cr}^*$. Although the quasi-static consideration is not fully justified anymore, the values of critical acoustic pressures or critical bubble sizes remain essentially valid for the sinusoidally driven bubble (if not too high driving frequencies are considered). However, different to the static case, any applied tension $-p_{ac} < 0$ will now change to overpressure $+p_{ac} > 0$ after half an acoustic cycle, and the otherwise, "infinite" expansion will be reverted toward a collapse.

---

[4]The unlimited expansion occurs theoretically in an unbounded liquid volume. In a real situation, the nucleus expansion will be stopped by boundary conditions, but it can reach a "macroscopic" bubble size.

1.9   Bubble Populations and Response Curves

The Blake threshold is crucial in acoustic cavitation since for higher driving pressures, it separates small "passive" bubbles (that oscillate only weakly) from the larger "active" ones (that undergo strong collapse). Indeed, the highest bubble expansion and compression ratios can occur directly beyond that threshold, as can be seen from bubble response diagrams like those given in Fig. 1.4. The top graphs show the quantity $(R_{max} - R_0)/R_0 = \alpha - 1$ for varying bubble sizes under constant driving frequency and three different fixed pressure amplitudes in water. The "low" ultrasonic frequency of 20 kHz on the left is contrasted to the "high"-frequency ultrasound of 500 kHz on the right. The form of response curves is very similar for both frequencies, and in particular the jump in relative expansion at the small radius side is apparent—the Blake threshold. The transition from small to rather large expansion ratios occurs very sharply for increasing R_0, and the bubbles of strongest relative expansion are only slightly larger than the largest passive ones. It is remarkable that still larger bubbles decrease again in their relative expansion (although considerable α values persist). In particular, the bubbles of resonant size, i.e., those with f_{res} matching the driving frequency f, do not peak significantly between the other bubble sizes for higher driving amplitudes. This fact is contradictory to former common notions that mainly resonant bubbles would be activated by ultrasound; one finds rather that usually much smaller bubbles, just beyond the Blake threshold, are main actors in acoustic cavitation at higher acoustic pressures.

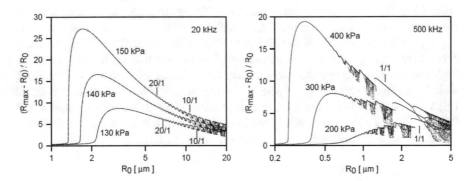

Fig. 1.4 Response curves of spherical bubbles driven at 20 kHz (left) and at 500 kHz (right). In each plot, the normalized maximum bubble expansion during one driving period, i.e. $(R_{max} - R_0)/R_0 = \alpha - 1$, is shown versus R_0. Three curves with indicated driving pressures are given for both frequencies. If higher periodic or aperiodic (chaotic) oscillations occur, the distinct maxima are plotted, which results in multiple points related to R_0 and partly in a gray-shaded broadening of the curves. Hysteresis (i.e., more than one stable solution) is included in the same way. Resonances are partly numbered: 1/1 refers to the linear resonance where $R_0 \sim R_{res}$, 10/1 and 20/1 indicate the nonlinear resonances at $R_0 \sim R_{res}/10$ and $R_0 \sim R_{res}/20$ (Keller–Miksis model, water at ambient conditions). Reproduced from [38], © Kluwer Academic 1999

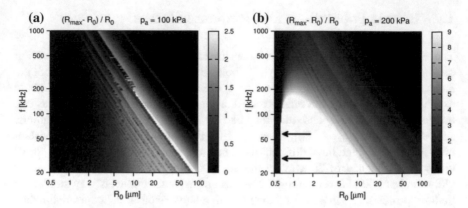

Fig. 1.5 Response diagrams of spherical bubbles driven at $p_a = 100$ kPa (**a**) and at 200 kPa (**b**). The normalized expansion $(R_{max} - R_0)/R_0 = \alpha - 1$ is shown grayscale coded versus rest radius R_0 and driving frequency f. In the right plot, $\alpha - 1$ was cut above 9 for better visibility of the structures. The arrows indicate the jump line connected with the Blake threshold of surface tension that occurs at $p_a = 200$ kPa above $R_0 \approx 0.6$ μm

A closer comparison of both diagrams reveals that the absolute values are shifted to smaller radii and higher pressure amplitudes for the higher frequency, which is a consequence of the faster driving oscillation with shorter tensile phase. A synopsis of spherical bubble response in a larger plane of driving frequency f and equilibrium radius R_0 is given in the graphs of Fig. 1.5, depicting the calculated and grayscale coded values of $\alpha - 1$. It can be seen that at the moderate driving pressure of 100 kPa (Fig. 1.5a), the strongest response for a given driving frequency occurs still near the respective linear resonance radius R_{res} which is connected with the frequency via $R_{res} \cdot f_{res} \approx 3$ (m/s) (a rule of thumb for water at ambient conditions). For the logarithmic axes, this relation shows up as a bright diagonal line running from small radii at high frequencies to larger radii at low frequencies. At the lower frequencies, however, already nonlinear resonances appear at smaller bubble sizes, manifesting itself in brighter stripes parallel to and to the left of the main diagonal. The higher amplitude of 200 kPa (Fig. 1.5b) reveals clearly the importance of the Blake threshold (arrows), and the linear resonance ceases to be a prominent feature.

1.10 Toward Realistic Bubble Systems

Solutions of spherical bubble models can give primary information on the behavior of driven bubbles in an ultrasonic field. Refinements of such models with respect to the interior gas and vapor, heat conduction, and chemical reactions can improve the validity of results obtained; see [27, 30, 31]. Now, we want to draw the attention onto the fact that in realistic active acoustic cavitation environments, like that shown in Fig. 1.6, an individual spherical bubble driven in a homogeneous sound

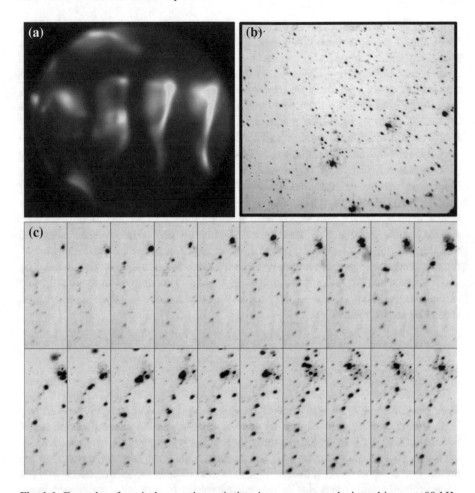

Fig. 1.6 Example of typical acoustic cavitation in a resonator device, driven at 90 kHz.
a Sonoluminescence image from an aqueous sodium chloride solution sparged with argon,
recorded by a long-term (30 s) exposure digital color photograph (color online) [46]. Frame width
is 5 cm. Due to the standing wave distribution of the sound field, bright stripes are visible.
Furthermore, parts of the emissions show orange color, while other parts appear blue. This
indicates differences in the bubble dynamics, possibly liquid injection into collapsing bubbles in
the orange regions (see Sect. 1.20). **b** Typical short-term exposure (1 μs) of a cavitating region
where sonoluminescence occurs (part of a high-speed series, frame width 2 mm). On this scale,
many individual bubbles of various smaller sizes are visible and a few larger bubbles that attract
and collect the smaller bubbles. **c** Detail of the high-speed series (turned 90°, size of each frame ca.
0.3 mm × 1 mm, time between frames 143 μs, exposure time 1 μs). Typical bubble velocities are
in the range of 10–20 cm/s. Bubbles appear small (near collapsed state) on the first frames and
then larger on the later frames (near maximum expansion state). This is result of a beat between
recording rate (7000 frames/s) and ultrasonic driving frequency (90,000 Hz). See also the
examples shown in [18]. Figures **a** and **b** reproduced from [47], © VCH Weinheim 2018

field and far from disturbances like other bubbles or boundaries is an exception. Most important deviations from this idealization are inherent instabilities of the spherical bubble shape, gradients in the acoustic driving field, interactions with neighboring bubbles and objects, and potentially a bulk liquid flow. Such factors can lead to bubble deformation and bubble movement in the liquid. Furthermore, bubble collisions and splitting are quite common in multibubble systems. This is illustrated in Fig. 1.6c where a sequence of a high-speed recording is reproduced frame by frame. Bubble motion and coalescence can be traced this way, and the very dynamic bubble life might be perceived, taking into account the fast timescales (the sequence in Fig. 1.6c covers 2.85 ms). Nevertheless, between collisions, the bubbles are more or less well separated, and many appear at least roughly as spherical entities. Thus, we discuss in the following several important aspects for multibubble systems again on basis of a single bubble. This has to be seen as a starting point, since a full description of a larger ensemble of oscillating and moving, merging, and splitting bubbles like in Fig. 1.6, under relevant inclusion of the modified sound propagation and induced turbulence, remains a very demanding and delicate task. Thus, the "bottom-up" approach [15] is seen as an applicable way to learn more about cavitating systems piece by piece, while other aspects might stay aside in the discussion.[5]

1.11   Spherical Stability

The spherical shape that is typically assumed by individual bubbles minimizes surface energy; i.e., it is caused by surface tension. This spherical shape might become unstable under volume oscillations and/or systematically deformed if the symmetry is externally perturbed, for instance, due to a relative motion of the bubble in the liquid or neighbored objects. In case of extreme deformations, the gas domain finally splits and the bubble disintegrates. Further down the *systematic* shape deformation of bubbles due to asymmetric flow fields and jetting will be considered. Now, the *instability* of spherical bubble shape is discussed. Shape instabilities can occur in principal without external trigger via strong amplification of minute disturbances. The instability of a plane boundary between heavier and lighter fluid when accelerated toward the heavier fluid is known as Rayleigh–Taylor instability [48]. In the case of a spherical bubble, it turns out that essentially a rapid collapse phase can be Rayleigh-Taylor unstable, while the expansion is stabilizing [8, 49–53]. Furthermore, a spherical ("breathing") mode of bubble oscillation can be accompanied by surface oscillation modes, as illustrated in Fig. 1.7 [8, 53–56]. The modes are ordered with an integer n that essentially

---

[5]In a way as a contrast, "top-down" descriptions of cavitation start from multiphase flow of liquid and vapor (for hydrodynamic cavitation, see [9, 39]) or from sound propagation in bubbly media (see [40–45]).

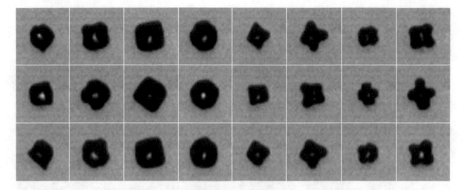

Fig. 1.7 Surface modes of a bubble moderately driven at 18 kHz in partly degassed water. Frames from left to right and top row to bottom row, frame height is about 500 μm, and the mean bubble radius is about 110 μm. Interframe time is 6.7 μs, and roughly, the eight frames of a row correspond to one acoustic driving period. Clearly, modal oscillations of $n = 4$ are visible with some addition of $n = 3$, and the bubble size indeed falls close to the resonance condition for the fourth mode stated in the text (image C. Cairós and A. Troia)

indicates the symmetry. Such modes lead to less focused collapse and potentially to ejected satellite bubbles. As well, droplets of liquid phase may be ejected into the bubble as a consequence of modal oscillations.

When unstable, surface modes can grow strongly in amplitude and finally disintegrate the bubble. Excitation of surface modes is a parametric resonance phenomenon and appears prominently for the smaller n and if the modal eigenfrequency f_n falls close to an integer multiple of half of the driving frequency f: $f_n = \sqrt{(n^2 - 1)(n + 2)\sigma/(\rho R^3)}/2\pi \approx m \cdot f/2, m = 1, 2, \ldots$. This estimate is valid only for small oscillations when $R \approx R_0$. In case of stronger volume oscillations of the bubble, deviations from this rule occur, as numerical calculations and experiments show [56]. Since the viscous damping is stronger for higher modal order, the modes of smaller n are generally the most unstable ones. For illustration, Fig. 1.8 shows calculated stability regions of the lowest three modes for driven bubbles in the parameter plane of driving amplitude and bubble equilibrium size. Note that the modal number n starts with $n = 2$ since $n = 0$ is the spherical (volume) mode, and $n = 1$ corresponds to bubble translation. The gross picture shows that both for smaller bubble sizes and for lower driving amplitudes, the surface modes become stable, while an increase of both quantities tends to destabilize the spherical shape. Closer inspection shows unstable peaks and stripes that reflect the resonances where the modal and the half driving frequency match as discussed above. Additionally, the expansion ratio is indicated by the lines of $\alpha = 2$ and $\alpha = 3$. Bubbles of high expansion ratios that remain stable occur only at the small radii and higher driving within a stripe close to the Blake threshold. Furthermore, it can be seen that larger bubbles become parametrically unstable already before a large volume expansion can be reached.

An important consequence of shape instabilities is the *limitation of active bubble sizes* in ultrasonically driven cavitation systems and thus of the reachable collapse

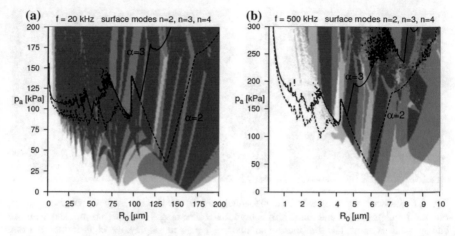

Fig. 1.8 Regions of parametric surface mode instabilities of a driven spherical bubble in water at 20 kHz (**a**) and 500 kHz (**b**), calculated with the equations given in [15]. The parameter plane of bubble equilibrium size R_0 and driving pressure amplitude p_a is colored according to instability of modes $n = 2$, $n = 3$, and $n = 4$: white: stable; light gray: one unstable mode; middle tone: two unstable modes; dark gray: three unstable modes. The lines indicate boundaries where the expansion ratio $\alpha = R_{max}/R_0$ reaches 2 (dashed line) and 3 (solid line). Above the lines, alpha gets larger. Zones of stable, strongly collapsing bubbles appear to the upper left

conditions. After nucleation, the bubbles may grow by gas diffusion and by merging events, but the final size will be determined by the instabilities. Only in regions of lower acoustic driving, e.g., near pressure nodes, the larger bubbles can stably exist. Since the instability threshold is crossed both for increasing bubble radius and for increasing pressure amplitude, in real systems, the energy concentration by the bubble collapse, reflected roughly by the expansion ratio α, cannot be amplified arbitrarily by an increase in driving. From Fig. 1.8, one reads that an "escape" from spherical shape instability appears most possible for very small bubbles. However, there the Blake threshold of surface tension can block bubble expansion (the lines $\alpha = 2$ and 3 are indicated). Furthermore, additional limitations in form of *gas diffusion* and *acoustic forces* occur. Both aspects are briefly outlined in the following to complete the discussion of system parameters where single spherical bubbles can stably exist (sometimes such a parameter region is termed "bubble habitat" [16]).

1.12 Gas Diffusion

For a complete discussion of the effects acting on the cavitation bubble population, gas diffusion has to be considered. The amount and type of dissolved gas in the liquid are relevant not only for the content of the bubbles, but also for their nucleation, growth, or dissolution. With respect to nucleation, typically governed by the nuclei present, it is expected that the dissolved gas concentration has an effect on them: It has been observed that cavitation inception is easier at higher concentrations of dissolved air and that degassing and pre-pressurizing lowers the cavitation threshold in terms of the tension needed for cavitation bubbles to occur [8]. A free small bubble of non-condensable gas should dissolve due to the Laplace overpressure $p_s(R) = 2\sigma/R$ inside, if the liquid is not supersaturated with the gas. This pressure will push out the gas into solution, and even more so for the smaller getting bubble, until the bubble vanishes. This is somehow not happening for the nuclei that are potentially fully or partly covered by hydrophobic substances, and several explanations are employed for this observation (static and dynamic ones; see, for instance, the discussion by Yasui et al. [22] and the related literature). Many of the explanations, however, include the background concentration of gas in the liquid. Thus, it is concluded that the amount of dissolved gas has—at least—an indirect influence on the nucleation of cavitation bubbles, namely via the stabilization of nuclei. In essence, the more gas dissolved, the more stable and abundant are the nuclei, and cavitation is facilitated.

Interestingly, the tendency of small bubbles to dissolve can be counteracted by volume oscillations. This phenomenon is known as *rectified diffusion* [7, 8, 57]. Since the gas diffusion through the bubble interface is proportional to the surface area, an expanded bubble has higher inflow per time than outflow in collapsed state. Furthermore, the gas concentration gradient across the bubble wall turns out to be higher in the expanded phase than in the contracted phase, which acts in the same direction. Additionally, nonlinear oscillations can result in longer times of large than of small radii—again supporting inflow against outflow of gas. These effects can outbalance the high Laplace pressure, and an acoustically driven bubble can finally accumulate gas from the liquid and grow in size if it is oscillating strongly enough. The boundary beyond which the bubbles grow is termed rectified diffusion threshold, and it can be crossed by increasing the driving pressure, the bubble size, or the liquid saturation level (for a more detailed quantitative discussion, see for instance [8, 58]). In conclusion, the rectified gas diffusion in driven bubble systems leads to a parameter regime where single bubbles grow in size, as opposed to the complementary parameter region where they dissolve. The rectified diffusion threshold thus constitutes a parameter boundary for active cavitation bubbles: Only beyond it, one expects relevant bubble populations. The growing bubbles, of course, will finally encounter shape instabilities and therefore be limited in size, as discussed above. Let us further note that for stronger driving pressures and small bubbles, the rectified diffusion boundary approaches the Blake threshold, as shown by Louisnard and Gomez [37]. Then, only active bubbles grow *by diffusion*. Another important mechanism of bubble growth is *by collision* with other bubbles. This is consequence of acoustic forces and discussed in the following.

1.13   Bjerknes Forces and Bubble Translation

Acoustic cavitation bubbles are typically moving. The reason behind this lies in acoustic forces, i.e., forces resulting from the interaction of bubbles and sound field. Such forces have been derived first by Bjerknes [59] and are nowadays called Bjerknes forces. An acoustic wave propagates variations of pressure p and density ρ in space via the oscillatory motion of the medium and thus is described by spatially and temporally varying quantities $\tilde{p}_{ac}(x, t)$, $\tilde{\rho}(x, t)$, and $\tilde{u}(x, t)$. The particle velocity is indicated by u, and the tilde indicates that the variations are around an equilibrium value p_0, ρ_0, and u_0 (the latter equals zero in the absence of a bulk flow). While the local *pressure* $p(x_b, t)$ at the position x_b of the bubble drives its volume pulsation, it is the *pressure gradient* $\nabla p(x_b, t)$ that is responsible for an acceleration of the bubble and therefore its spatial motion. Summing up the pressure forces over the bubble wall leads to the simple expression $F(t) = -V(t) \cdot \nabla p(x_b, t)$ as the resulting momentary net force on the bubble of volume $V(t)$ if the pressure variation on the scale of the bubble size can be assumed linear (long acoustic wavelength approximation: $R \ll \lambda = c/f$) [7, 8]. A simple application is the buoyancy force acting in gravity. Then, $\nabla p = \rho g$ is the (constant) hydrostatic pressure gradient in a liquid of density ρ, g being the gravitational acceleration acting downwards. The resulting buoyancy force is directed upwards and results as $F_{buoy} = -V\rho g$. In acoustic fields, both pressure and pressure gradient are oscillating, and also the bubble volume varies with time.[6] Thus, the time average of $F(t)$ is considered which is termed *primary Bjerknes force*: $F_{B1} = \langle -V(t) \cdot \nabla p_{ac}(x_b, t) \rangle_T$. The temporal average is denoted by the brackets $\langle \cdot \rangle_T$ and is taken over the driving period $T = 1/f$ since a periodic variation with this period is assumed (otherwise, longer time averages have to be taken). It is important to note that this average and thus the Bjerknes force does typically *not* vanish, even if the gradient (and possibly also the bubble volume variation) is a sinusoidal function of time with zero time average. The reason is that bubble volume and gradient do not oscillate independently, but rather with a fixed phase relation. The textbook case considers a linearized bubble oscillation $R(t) = R_0[1 + \epsilon \sin(\omega t + \phi_0)]$ driven by small amplitude harmonic plane waves, leading to analytical expressions (see for instance [7, 8]). The result is that for a plane standing acoustic wave field $p_{ac}(x, t) = p_a \cos(kx) \cos(\omega t)$ with the wave number $k = \omega/c$, the bubbles larger than the linear resonance size R_{res} are forced toward the pressure nodes, while bubbles smaller than R_{res} are pushed toward the pressure antinodes. In plane traveling waves of the form $p_{ac}(x, t) = p_a \cos(\omega t - kx)$, all bubbles are pushed forwards (in $+x$-direction), and those close to R_{res} feel the largest force. Important modifications of this picture appear for the case of higher pressure amplitudes that lead to larger, non-sinusoidal bubble volume excursions and stronger collapse oscillations. Figure 1.9 shows an illustration

---

[6]Pressure gradients of the sound field are typically much larger than the hydrostatic pressure gradient, and therefore buoyancy can often be neglected in the discussion of acoustic cavitation bubbles. Only for larger bubbles and weak driving, buoyancy might supersede acoustic forces which leads to a rise of the bubble to the surface.

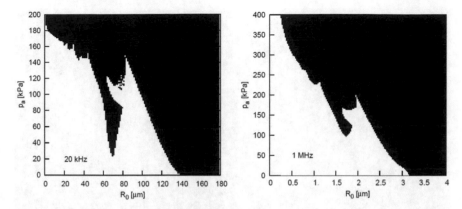

Fig. 1.9 Primary Bjerknes forces in a standing acoustic wave of 20 kHz (left) and 1 MHz (right). The colors show the direction of the force for variation of sound amplitude p_a and bubble equilibrium radius R_0: White indicates a force toward the high pressure regions, dark areas a force toward the pressure nodes. The regions are separated on the R_0-axis at the resonance radii ($R_{res} \approx 140$ μm for 20 kHz and $R_{res} \approx 3.2$ μm for 1 MHz; water at normal conditions, isothermal calculation with $\kappa = 1$). Reproduced from [15], © Universitätsverlag Göttingen 2007

of primary Bjerknes force direction including higher p_a in a standing wave field. Notably, also bubbles smaller than resonance size experience a *repulsive* force away from the antinode at sufficiently high pressure amplitudes ("Bjerknes force reversal" [60, 61]). This is an additional phenomenon that limits access to very strongly driven bubble collapse: Exceedingly expanded bubbles tend to leave the high pressure zones in standing waves.

The secondary Bjerknes force $\boldsymbol{F}_{B2}$ is similar to the primary one, but results from the scattered sound field p_{sc} from a neighbored bubble instead of the incident sound field p_{ac}: $\boldsymbol{F}_{B2} = \langle -V(t) \cdot \nabla p_{sc}(\boldsymbol{x}_b, t) \rangle_T$. To a certain approximation, the force of bubble 2 at position $\boldsymbol{x}_{b2}$ exerted on bubble 1 at location $\boldsymbol{x}_{b1}$ can be expressed as $\boldsymbol{F}_{B2} = 4\pi\rho \langle \dot{V}_1(t) \cdot \dot{V}_2(t) \rangle_T (\boldsymbol{x}_{b1} - \boldsymbol{x}_{b2})/d^3$ which means that its strength decays with the reciprocal squared bubble–bubble distance $d = |\boldsymbol{x}_{b1} - \boldsymbol{x}_{b2}|$ [62]. If sizes and oscillations of the neighbored bubbles are similar, the force leads typically to mutual attraction. Only for bubbles oscillating in antiphase (like one larger and one smaller than the resonance size), the force is repulsive [8, 63]. Again, nonlinear bubble oscillations in the stronger driving regime as well as coupling between the bubbles generate more complicated and often larger secondary Bjerknes forces [62]. In the most cases, however, the secondary Bjerknes forces lead to strong attraction of adjacent bubbles and finally to coalescence. Figure 1.10 is illustrating such a case for two bubbles that oscillate such strongly that sonoluminescence flashes occur in their collapse. The blurry darker silhouettes track the bubble maximum size, while the flashes mark the points of bubble collapse. Both outline the bubble paths, and the bending of the trajectories toward each other as well as the strong bubble acceleration right before coalescence can be perceived.

Figure 1.11 illustrates the partly complicated behavior of the secondary Bjerknes force at higher driving pressures. At low driving amplitudes, the parameter plane of

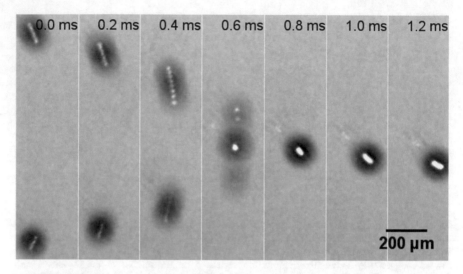

Fig. 1.10 Collision of two sonoluminescing bubbles under the action of the secondary Bjerknes force. After merging, the flashes continue and get brighter (xenon bubble in phosphoric acid driven at 36.5 kHz, recording at 5000 frames/s). Reproduced Fig. 2a from [64], © American Physical Society

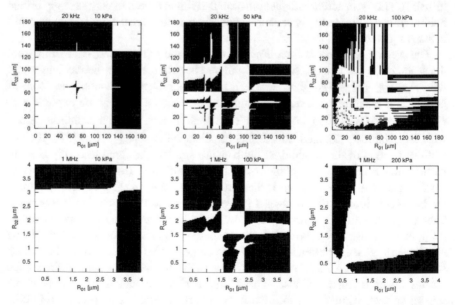

Fig. 1.11 Sign distribution of the secondary Bjerknes force between two bubbles of equilibrium radius R_{01} and R_{02} for fixed pressure amplitude p_a. White areas indicate an attractive force between the bubbles, and dark areas indicate mutual repulsion. Top row: $f = 20$ kHz, $p_a = 10$, 50, and 200 kPa; bottom row: $f = 1$ MHz, $p_a = 10$, 100, and 200 kPa (R_{res} as in Fig. 1.9; water under normal conditions; $\kappa = 1$). Reproduced from [15], © Universitätsverlag Göttingen 2007

both bubble sizes is well divided into attraction and repulsion rectangles, as described by the textbook case of linear bubble oscillations [8]. For higher driving, the nonlinear resonances of the bubble oscillation lead to parallel (self-similar) repulsive stripes, since rather different oscillation phases are found as well around these resonances. At even higher excitation, additionally, the delay in collapse time at higher nonlinear expansion comes into play: The pattern returns to roughly the initial structure, but now with the Blake threshold radius playing the role of the linear resonance radius.

1.14 Viscous Drag Force

The moving bubbles experience a counteracting force due to viscosity of the liquid, the drag force F_D. For this discussion, one usually employs the Reynolds number $Re = R|u_b|\rho/\mu$. From measurements at cavitating systems like that shown in Fig. 1.6, one obtains typical bubble speeds of cm/s up to several m/s and typical (average or maximum) bubble radii in the range of a few up to about 100 µm. For water ($\mu = 10^{-3}$ Pa s), this means that Reynolds numbers up to a few hundred have to be considered.

Supposing first a non-oscillating bubble with a bubble centroid velocity u_b relative to the fluid, one finds for small Reynolds numbers Re the drag force $F_D = -4\pi\mu Ru_b$ [9]. Here, it is assumed that the bubble surface is "free" in the sense that the fluid molecules do not encounter shear stress. If the bubble is covered with a hydrophobic substance, a no-slip boundary condition would be more appropriate at the bubble wall, and the drag force at small Re increases to $F_D = -6\pi\mu Ru_b$. This is the classical result of Stokes' drag of a moving sphere in fluids. In many cases, bubbles in real liquids (in particular water) are better described by this no-slip formula [9, 65]. This fact is apparently due to frequently given hydrophobic contaminations. The drag on stationary bubbles at higher Re becomes more involved, and corrections to the given expressions occur. For example, Brennen [9] cites Klyachko [66] for a formula that fits data well up to $Re \approx 1000$: $F_D = -6\pi\mu Ru_b[1 + Re^{2/3}/6]$. If we now consider moving and oscillating bubbles, the analysis becomes even more complicated, and one partly has to rely on numerical calculations. An extended discussion has been given by Magnaudet and Legendre [67] with a derivation of low and high Re limits, the latter being $F_D(t) = -12\pi\mu R(t)u_b$. It is noted by the authors that this expression also applies for lower Re if the bubble wall velocity is sufficiently high, since the viscous dissipation is then dominated by the oscillation (vs. the translation). Thus, the formula is a good starting point to model the drag on strongly oscillating and faster-moving cavitation bubbles, and it has been shown to work well for experimental data [68].

1.15   Phase Diagrams

Going back to the parameter regions where active bubbles are to be expected in
ultrasonically driven cavitating systems, we put together the boundaries of shape
stability (SI), rectified diffusion (RD), and Bjerknes force reversal (BJ). For a plane
standing wave, one obtains a synopsis like those presented in Fig. 1.12, shown
there for 20 kHz and 1 MHz driving frequency (compare also the diagrams given
by Apfel [69] or Church [70]). In the plane of bubble radius R_0 and driving pressure
amplitude p_a, the corresponding lines are depicted. The SI boundary is shown only
for $n = 2$, and the RD threshold is given for two different levels of relative gas
saturation (1.0 corresponds to full saturation at ambient pressure of 1 atm, 0.1 to
only 10% saturation). Circles indicate points of accumulation for the non-degassed
water, i.e., values of stable bubble sizes and driving pressures that are reached under
diffusional growth and the action of the primary Bjerknes force. Of course, not all
bubbles will finally have the indicated radius and be located at a position of the
indicated driving pressure amplitude, since the real cavitating liquid constitutes a
dynamic multibubble system of translating, interacting, merging, and splitting
bubbles. However, the values are a reasonable estimation of an "average" bubble in
the system. For instance, the 20 kHz case suggests equilibrium radii in the range of
5 μm, and the measured distributions in Fig. 1.1b (there at 25 kHz) essentially
spread around this value. Note that this result is not trivial, although the nuclei
distribution shown in Fig. 1.1a for filtered tap water gives a similar size range. The
nuclei statistics is obtained under silent and static conditions, while the cavitation
bubble equilibrium sizes are result of complicated bubble and sound field interaction, as described above. Thus, the bubble population of a cavitating system is
based on a dynamical process. Often, it is justified to assume a stationary situation
of this dynamical system which allows a quasi-static analysis, as proposed in the
phase diagrams of Fig. 1.12. Note further that non-active bubbles of larger size,

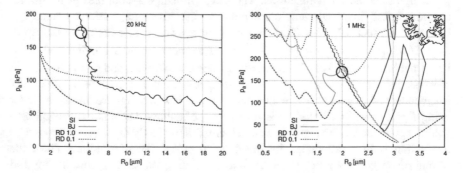

Fig. 1.12 Phase diagrams for bubbles in a standing wave, indicating the lines of surface
instability (SI) and Bjerknes force inversion (BJ). Also, the thresholds of rectified diffusion are
given for water at gas saturation (RD 1.0) and degassed to 10% of saturation (RD 0.1). Left:
$f = 20$ kHz, right: $f = 1$ MHz (water at normal conditions, $\kappa = 1$). Points of accumulation are
marked by a circle. Reproduced from [15], © Universitätsverlag Göttingen 2007

complying $R_0 > R_{res}$, can be trapped in low driving pressure zones, e.g., near standing wave antinodes. Their parameter region in the phase diagram stretches along the $p_a = 0$ axis from the linear resonance radius on to the right: beyond $R_0 \approx 140$ μm for 20 kHz (not visible in Fig. 1.11) and beyond $R_0 \approx 3.2$ μm for 1 MHz. Since rectified diffusion into the bubble works only for larger volume pulsations, these weakly oscillating bubbles tend to dissolve due to surface tension. However, they can collect gas through merging processes with other bubbles that are driven to the antinodes as well and grow nevertheless. Then, these passive bubbles can rise due to buoyancy once large enough to overcome the trapping primary Bjerknes forces.

1.16 Structure Formation

From the discussion of secondary Bjerknes forces above, it becomes clear that a spatially homogeneous distribution of acoustic cavitation bubbles should be unstable. Since similar neighboring bubbles attract each other, they will develop clusters, somehow like stars in the universe form galaxies.[7] Furthermore, the spatial distribution depends on the characteristics of the driving ultrasonic field. In particular, spatial modulations of pressure amplitude, as found in standing or decaying wave fields, will cause bubble motion on scales of the acoustic wavelength via primary Bjerknes forces and potentially separation of larger and smaller bubble sizes. Indeed, acoustically cavitating systems usually develop bubble structures, and specific patterns of spatial bubble locations are found. A variety of them has been described in [71], and here, we show just a few typical ones in Fig. 1.13. Within a structure, the bubbles undergo frequent collisions, merging, and splitting processes, but overall the multibubble system stays in a stationary state that is recognizable as a certain structure. In clustering patterns, strongly oscillating bubbles travel inwards, attracted first by primary and later by secondary Bjerknes forces toward other, already accumulated bubbles. After coalescence and growth, they become shape unstable and split off tiny bubbles that remain passive due to their surface tension pressure. These small, very weakly oscillating bubbles feel low Bjerknes forces and are transported outside the structure again by liquid convection. The liquid flow, which is considered as turbulent in and near the bubble agglomerates, thus controls the motion and redistribution of these microbubbles. They are occasionally perceived as a misty cloud around the bubble structure. Later, they dissolve or serve as nuclei again.

---

[7]While secondary Bjerknes forces indeed decay with the squared distance like gravitational forces, there are differences in that stars move without friction and do typically not collide. Furthermore, the secondary Bjerknes force changes for very close or far distances, and the "mass" of a bubble depends on the driving pressure at its position. Nevertheless, partly interesting similarities exist visually between bubble structures and galactic structures.

Fig. 1.13 Some typical cavitation bubble structures generated by strong ultrasonic fields in the frequency range of 20–40 kHz. **a** Double layers around nodal planes of a horizontal standing wave field. **b** Filament in a traveling wave field in front of a transducer. **c** Snapshot detail of a filament. **d** Small bubble cluster at different phases of bubble oscillation: nearly collapsed at 8 ms, expanded at 9 ms, intermediate bubble sizes at 10 ms. **e** Flare structure near a submerged transducer (placed to the left) (Images **c** and **d** from [15], © Universitätsverlag Göttingen 2007)

1.17   Bubble Traps

While the bubble structures as a whole are quite stationary, the individual bubbles within the structure are not. They are normally transient in space and time, and thus, observations of details of their dynamics over a longer time interval are difficult. This means that many theoretical considerations and models are hard to verify on a single-bubble level, in particular the large radial oscillations and the rapid hard collapse. However, it is possible to isolate strongly driven bubbles and drive them under stationary conditions in so-called *bubble traps*. Such devices use a non-cavitating acoustic standing wave field in a resonator of one or few acoustic wavelengths size. Then, a single bubble is seeded that runs toward the pressure antinode, where it is caught. If the bubble is small enough, primary Bjerknes forces will ensure a capture against buoyancy. The peculiar point is here that the bubble permanently performs extreme volume pulsations with strong collapses and can show light emission—*sonoluminescence*—and chemistry—*sonochemistry*.[8] An example for a rectangular resonator cell and the observed bubble dynamics of the trapped, light-emitting bubble are shown in Fig. 1.14. Other geometries like cylinders or spheres work as well. Seeding can be accomplished by electrolysis, by a focused laser pulse, or by air entrainment of a droplet falling onto the liquid surface. The liquid is usually degassed to avoid bulk cavitation, but also to provide for a diffusionally stable trapped bubble that otherwise would grow by rectified diffusion

---

[8]Inactive larger bubbles can be trapped at pressure nodes of a standing acoustic wave.

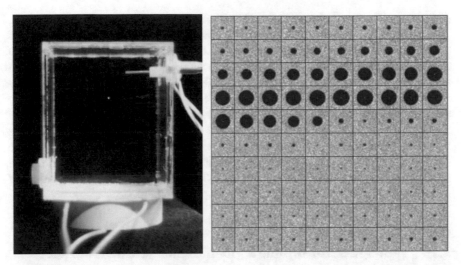

Fig. 1.14 *Left*: Rectangular bubble trap with a piezoelectric transducer glued to the bottom (cell width 5 cm). The bluish-white spot in the middle of the upper part of the water-filled cuvette is the light emitted by a stably oscillating bubble, visible on this photograph of 20 min exposure. *Right*: Series of photographic short-term exposures of a similar trapped sonoluminescing bubble (line by line, time between frames 500 ns, frame size 160 μm × 160 μm). The bubble appears dark in front of the bright background. Radius–time dynamics correspond roughly to the curve of 130 kPa in Fig. 1.3a, and the according large volume variations can be perceived. Driving frequency is 21.4 kHz, and thus, the 100 frames cover roughly one full acoustic cycle (both images courtesy of R. Geisler, see also [16])

and rise. It turns out that under variation of the main external parameters, namely acoustic pressure amplitude, dissolved gas type and content, and ambient pressure, several stable regimes of bubble dynamics can be reached and observed (the driving frequency is usually fixed by the trap's geometry). Many important findings with respect to sonoluminescence (SL) have been accomplished in bubble traps, the phenomenon then termed *single-bubble sonoluminescence* (SBSL). See, for instance, the seminal article by Gaitan et al. [72], experiments and discussions by Putterman and coworkers [73, 74] and the rich literature in Young's book [10] and Crum's resource paper [75]. Results comprise the determination of sonoluminescence flash duration [76], spectral time traces of the flash [77], and many substance and parameter studies on brightness and emission spectra [78, 79], including evidence of a plasma in the collapsing bubble [80, 81]. Further, important experiments with "levitated" (trapped) bubbles consider the observation of chemistry from a single bubble [33, 82–85] and bubble dynamics and stability studies [55]. Also, the Bjerknes force reversal has been shown experimentally in a bubble trap [86, 87].

Bubble traps constitute a unique environment to prepare stable bubble regimes with extreme collapse phenomena, and they are ideally suited for experimental observations. Several results discussed later in Chap. 2 rely on trapped bubbles. On the other hand, the traps are usually designed for individual bubbles, and phenomena like translation, coalescence, or splitting are intentionally not captured. Thus, it remains to be explored how far the captured bubbles represent members of the multibubble environments. For instance, it is well known that optical emission spectra can differ between multibubble sonoluminescence systems (MBSL) and single SBSL bubbles. In particular, spectral features like prominent emission lines are partly absent under SBSL conditions [88], most probably linked with the absence of disturbances and instabilities that are present in MBSL environments. Some aspects thereof are presented in the following. There is still a bridge to close between the isolated active bubbles and a large bubble ensemble like in the structures illustrated above. Experiments with few-bubble systems or individual unstable bubbles might offer advances here [33], and this question remains a field of active research.

1.18   Non-spherical Bubble Dynamics and Jetting Collapse

Above, we have treated instabilities of the spherical shape of an oscillating bubble in form of unstable surface modes. These are triggered by incidental events from pervasive fluctuations, and the exact bubble shape dynamics has a random or non-reproducible component. A different source of non-spherical bubble dynamics is given by an asymmetric environment of the bubble: Any deviation from isotropic geometry can cause a systematic and reproducible bubble deformation by imposing a gradient in pressure. Relevant cases are adjacent boundaries from walls, particles, other bubbles, or a free surface. Furthermore, the time-varying pressure gradient of the sound field and a hydrostatic pressure gradient due to gravity (buoyancy) will disturb the symmetry. Last but not least, the pure motion of the bubble relative to the liquid represents an asymmetry. However, an answer to the question to what extent an initially spherical bubble shape will be significantly affected under the asymmetries is not straightforward. It will depend on the bubble size, the relative strength of disturbance, on the oscillatory bubble motion and on parameters like surface tension and viscosity. For example, a non-oscillating bubble rising under buoyancy deforms into an oblate ellipsoid at small velocities and into a spherical cap at higher speeds [9]. An expanding and collapsing bubble under translation can develop a *liquid jet* in direction of the relative motion [89]. This is illustrated in Fig. 1.15 and described in more detail below. The jet is a peculiar phenomenon of collapsing cavities that significantly affects the bubble and its vicinity: By piercing the bubble and hitting the opposite bubble wall, the liquid jet changes the bubble topology from a sphere to a torus. The subsequent bubble collapse therefore proceeds toward a ring, not toward a point. The interior gas phase of a jetting bubble becomes less compressed than in a spherical collapse, and peak temperatures are lower to some extent. One reason is that part of the collapse energy goes to liquid kinetic energy of the jet flow and not to gas compression [90]. If one assumes an inner structure of the gas like ingoing focusing waves or shocks [91, 92], such temperature enhancing mechanisms would also work less in a non-spherical

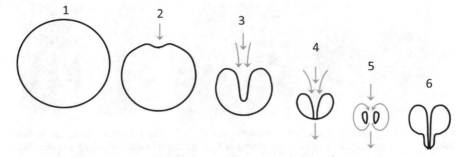

Fig. 1.15 Schematic sketch of a jetting bubble collapse induced by relative motion of bubble and liquid (not to scale; variants of shapes and timings are possible). The lines indicate the bubble wall in an axisymmetric cross section; the arrows represent liquid flow connected with the jet. For simplicity, the inward flow during collapse (1–5) and outward flow during rebound (6) are not shown. The jet hits the opposite bubble wall at instant (4), and the afterward torus-like bubble collapses further (5). In the re-expansion (6), the jet flow persists and entrained gas forms a characteristic bulge or peak ("nose")

collapse. Additionally, the jetting process can lead to liquid phase being injected into the heated gas which would have a cooling effect, but as well has impact on potential chemical reactions [34, 93, 94].[9] Furthermore, circulation is introduced in the liquid [89, 96], which manifests itself in vortex flows around the bubble [97] and small-scale turbulence. This is important for mixing and transport processes in the liquid and can also increase shear forces exerted by the bubble oscillation [98].

1.19 Jetting at Solid Objects

Similar jets occur for bubbles collapsing close to a solid boundary, for example, near objects [16, 99, 100]; see Fig. 1.16. Then, the jet develops toward the boundary, and the bubble also approaches the object. Jet impact and flow, bubble gas phase, and the induced vortex flows can directly interact with the boundary, which is particularly relevant for erosion of material by cavitation [101] and for ultrasonic cleaning applications [5, 102–104]. In both cases, bubbles collapsing in direct vicinity of the surfaces are responsible for the observed effects of damage or dirt removal [101, 105]. The detailed mechanisms comprise jet impact and shock waves for material damage, and additionally shear forces and sweeping of the

---

[9]Details of liquid injection are still subject of investigation. At least, three scenarios could take place: (I) During re-expansion of the bubble, the spherical shape is roughly restored, and remnants of the jet might disintegrate into droplets, remaining in the gas phase until the next collapse happens. (II) The jet impact onto the opposite bubble wall can cause nanosplashes that disintegrate into droplets [95]. (III) The rear side of the bubble might become unstable and split off droplets. In this context, note the non-smooth bubble backside in Fig. 1.17.

Fig. 1.16 Jetting bubble near a solid boundary for $D^* \approx 1.4$. The boundary is at the bottom, visible via the reflections. The black silhouette is from an experimental recording where a laser pulse-seeded bubble collapses. The lighter gray (online: blue) part corresponds to the central cross section of an axisymmetric numerical calculation with the finite volume/volume of fluids method (see Koch et al. [112]). Due to the cut through the bubble, its interior toroidal structure is visible in the simulated data. The dark feature appearing atop of the experimental bubble after collapse consists of secondary cavitation bubbles. They are induced by the toroidal collapse shock wave ("counter jet" [113]) and is not captured by the simulations

contact line (gas/liquid/solid phase boundary) for cleaning. Jets occurring at solid interfaces are very well investigated due to their technical relevance, and a large literature exists on experiments and modeling (see, for instance, [14, 101, 106, 107] as a starting point). The main parameter for the dynamics is the normalized standoff distance $D^* = D/R_{max}$ where D is the distance between the center of the expanded bubble and the solid wall, and R_{max} is the maximum bubble radius before collapse. Although the bubble behavior scales rather well with D^* in a large range of parameters and irrespective of the real bubble size, the phenomena are still very complex, since the flows depend partly sensitively on the initial conditions [108]. Therefore, many details of bubble collapses at walls are still under investigation, and interesting findings could be reported recently with improving experimental and numerical tools. For instance, the vortex flow induced by the jet is inverted if the bubble collapses relatively close to the wall, i.e., D^* being smaller than about 1.35 [97]. Other work is concerned with details of jet impact and thereby produced secondary "nanojets" that can cause liquid transport into the gas phase [95]. Further actual research in this field deals with scaling approaches [109], more complicated surface geometries [110], or jets in microchannels [111] where viscosity and surface tension have to be taken into account.

1.20   Translation-Induced Jetting

The jet developed in the collapse of a moving bubble can be understood as a consequence of conservation of momentum [89]. If we consider a one-dimensional translation of a spherical bubble, the bubble position $x(t)$ can be approximately described by the following equation that is solved simultaneously with the RP or Keller–Miksis equation (1.4), (1.5) given above:

$$m_b \ddot{x} + \frac{2\pi}{3}\rho\frac{\mathrm{d}}{\mathrm{d}t}\left(R^3\dot{x}\right) = -\frac{4\pi}{3}R^3\frac{\mathrm{d}}{\mathrm{d}x}p_\infty(x,t) + F_d(R,u_b).$$

Here, m_b is the gas mass in the bubble and F_d is a viscous drag force as described before with the relative bubble velocity $u_b = \dot{x} - u(x)$. The term $\mathrm{d}p_\infty(x,t)/\mathrm{d}x$ is the pressure gradient that invokes the translational motion. Its value is taken at the bubble position as if the bubble would be absent, as is the value of the liquid velocity $u(x)$. This means that the acoustic wavelength should be much larger than the bubble size, similar to the Bjerknes force formulas given above. Figure 1.3b in Sect. 1.6 exemplifies bubble motion for increasing the driving pressure amplitude (which as well increases the pressure gradient amplitude) in a plane traveling wave running in positive x-direction. It is seen that weakly oscillating bubbles just gently move forth and back, while a stronger collapse leads to a significant acceleration and a forward jump of the bubble. This can be understood by momentary conservation of the *Kelvin impulse*. The Kelvin impulse (or *quasi-momentum*) is a quantity related (but not identical) to the momentum of the liquid displaced by the moving bubble, and it is relevant for any motion of bodies through a fluid [114, 115]. In our case, we can assume that the product of relative bubble center velocity u_b and *virtual mass* m_v of the bubble is approximately conserved. The virtual mass (also *added mass*) represents inertia of the flowing liquid and adds to the gas mass. For a spherical body like the expanded bubble, the virtual mass corresponds to half of the displaced liquid mass, $m_v = 2\pi\rho R^3/3$, and thus is proportional to the bubble volume [9]. Since the liquid density is much higher than the gas density (apart from the ultimate collapse peak when the gas is extremely compressed), the virtual mass is the dominant part responsible for the moving bubble's inertia. During implosion, the bubble volume and therefore the virtual mass shrink, and the bubble velocity has to speed up accordingly (roughly by R^{-3}). This acceleration becomes prominent for a strong collapse and is clearly visible in Fig. 1.3b as the jump of the strongest driven bubble at the main collapse (further leaps appear at the subsequent after bounce collapses). For very strong volume shrinkage and fast acceleration, the Kelvin impulse cannot be sustained by the accelerated virtual mass of the spherical bubble alone, and the forward liquid jet flow through the bubble occurs [89]. This jet is faster than the bubble center velocity and carries significant momentum, thus overtaking (and slowing down) the collapsed bubble [115, 116]. Kelvin impulse is transferred into a circulation (vortex flow) after the jet impacts the other bubble side, and thus, it can be conserved by the topological change to a torus bubble [89]. Of course, spherical bubble models cease to be valid at this moment, but the essential bubble motion until the jetting is captured quite well by RP-like models, as exemplified in Fig. 1.3b. Analysis of the detailed flow dynamics around jetting bubbles, as well as the interior gas heating and potential liquid injection, requires much more sophisticated models, and such questions are topic of actual research.

Some experimental images of jetting bubbles in ultrasonic fields are presented in Fig. 1.17. These bubbles, far from boundaries, are moving relatively fast due to primary Bjerknes forces, and the jet is induced by the described mechanism. It is

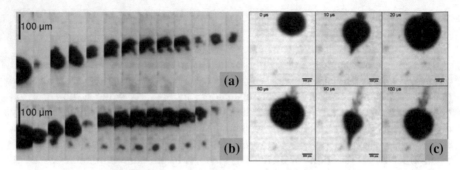

Fig. 1.17 High-speed recordings of moving and jetting acoustic cavitation bubbles. Series **a** and **b** show air bubbles in water below an ultrasonic horn, driven at 20 kHz (from [117], recordings at 250,000 frames/s). Collapse and jet take place between frames 1 and 3 (from left). The re-expansion of the bubbles is rather unstable, and shape deformations and bubble split-off are triggered by the jet. Series **c** shows selected frames of a high-speed recording of a xenon bubble in sulfuric acid (at 100,000 frames/s). The rebound is much more stable here, probably due to the high viscosity of the acid. On the other hand, the rear side of the bubble ejects microbubbles on its path [34], a feature not occurring in water here. Gas ejections on the backside of moving bubbles have, however, been observed for bubbles in argon sparged aqueous NaCl solution; see [64]. Figures **a** and **b** reproduced from [117], © American Physical Society 2014. Figure **c** reproduced from [34], © Elsevier 2017

important to note that acoustic cavitation bubbles are virtually always moving and thus are prone to undergo jetting collapses. However, the parameter thresholds for the transition from a still spherical collapse (in the sense of simply connected gas volume) to a torus collapse are not easily found, and one relies on numerical work. Resonant bubble sizes and/or increased driving amplitude will lead to larger expansion ratios and higher bubble velocities, finally causing a jet. For a traveling wave, Calvisi et al. show some results based on the boundary integral method, and it can be seen that indeed at sufficiently high driving, a jetting collapse will always develop [90]. In the context of sonoluminescence of different colors at different regions of the cavitation cloud (see [93, 118] and compare Fig. 1.6a), there are strong indications that bubble populations with and without jetting collapse are causing the difference [34]: Jetting can inject liquid that contains non-volatile components that are in turn responsible for the change of color.

1.21 Activity Mapping

We have seen that an acoustically cavitating system is composed of a distribution of bubbles, and not all are active in a specific, desired sense. For instance, sonoluminescence measurements from a bubble cloud will always monitor emission averaged over all emitters, and it remains unclear if spectral features are shared by the full population or just come from relatively few but relatively strong sources.

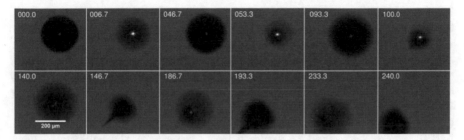

Fig. 1.18 Collapse events and light emissions of an accelerating, relatively large bubble of xenon in phosphoric acid. The bubble starts to develop stronger and stronger jetting during collapse, and the light emission ceases (driving at 23 kHz, 150,000 frames/s, timing given in µs, frame width 470 µm). Reproduced from [64], © American Physical Society 2017

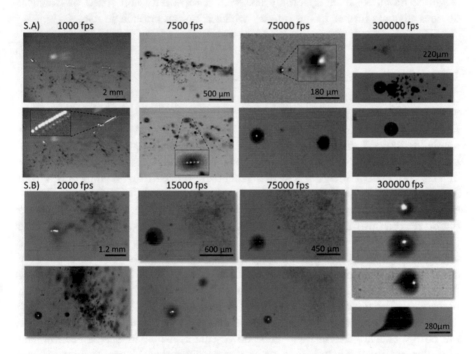

Fig. 1.19 Sample images from high-speed recordings at different frame rates and magnifications (xenon in phosphoric acid; top rows: 36.5 kHz; bottom rows: 23 kHz). Scales and recording speeds indicated. It can be seen that some moving bubbles emit light in every collapse and also that sonoluminescence can occur together with pronounced jetting. Reproduced from [64], © American Physical Society 2017

The same holds in principle for chemical reaction yields. Ideally, one would like to resolve emissions or reactions in space and time and relate activity to specific bubbles and their dynamics. This is yet not really possible in multibubble systems since the light (or yield) is usually much too weak to be detected in the timespan of a bubble oscillation or even bubble lifetime. Only in the case of the extremely bright SL emission from xenon bubbles in sulfuric or phosphoric acid, a first step in this direction has been demonstrated. By simultaneous high-speed monitoring of bubble dynamics and light flashes, various aspects of light-emitting bubbles in MBSL environment could be explored [64], for instance, the collision of luminescing bubbles, as shown in Fig. 1.10, or the dependence of flash brightness on jetting strength, given in Fig. 1.18. A collection of different scenarios, observed at a variety of exposure times, is presented in Fig. 1.19. As a perspective, a spatial mapping of bubbles and their light emissions within a larger cloud or structure might be provided, even spectrally resolved, if sensitivity and speed of technical devices advance further in the future—which is not an unrealistic idea.

1.22   Beyond Adiabatic Compression

The most interesting part of bubble dynamics—the collapse—is the most difficult to observe in direct imaging. Since scales are extremely fast and close to optical resolution limits, further advancing steps in imaging techniques and in technology have to be taken. Apart from an improvement of speed, sensitivity, and resolution of standard optical imaging sensors, potentially as well novel approaches with X-ray and electron microscope imaging are promising candidates. However, in any case, a lot of information about the conditions in the collapsing bubbles can be obtained from other measures: light emission spectra and chemical reactions. These topics are treated in the following chapters, and anticipating the results there, we will learn that the phenomena during the hot collapse phase can be even more complex and intriguing than the rest of the bubble life. In our discussion of bubble dynamics up to now, we did not go beyond a rather simple interior bubble model, namely an adiabatically and homogeneously heated ideal gas. For many aspects like shape instabilities or acoustic forces, this does not mean a severe shortcoming, since the energy exchanged via light emission or chemical reactions is rather low and often has no dramatic influence on the gross dynamics. However, if a realistic picture of the few extreme nanoseconds is required, one has to go beyond adiabatic heating. We have mentioned above models that include heat conduction, evaporation and condensation, and chemical reactions. More information on such models that employ ordinary differential equations is contained, for instance, in Yasui's booklet within this series [119]. Another method that by construction captures all these additional issues is a molecular dynamics simulation. An example from [92] is

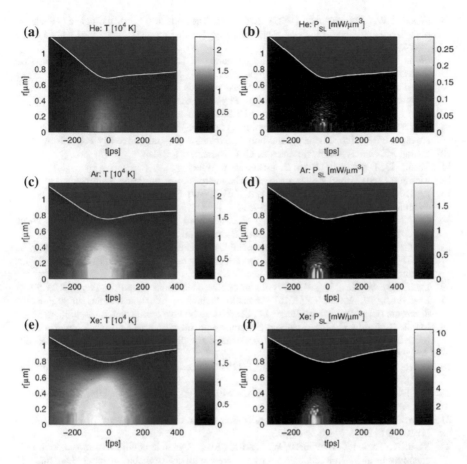

Fig. 1.20 Results from a molecular dynamics simulation [92]. The images show the evolution in space and time of temperature and light emission power in the case of helium (**a** and **b**), argon (**c** and **d**), and xenon (**e** and **f**) including vapor chemistry. One million hard sphere particles in a collapsing SBSL bubble of $R_0 = 4.5$ μm, driven at 26.5 kHz. Reproduced from [92], IOP Science 2012, CC-BY 3.0

shown in Fig. 1.20. In the future, more information from such quite expensive calculations and other advanced models is to be expected, giving insight into the otherwise unresolved hot spot of collapsing bubbles.

References

1. Silberrad D (1912) Propeller erosion. Engineering 33:33–35
2. Rayleigh L (1917) On the pressure developed in a liquid during the collapse of a spherical cavity. Phil Mag Ser 6(34):94–98

3. Wood RW, Loomis AL (1927) XXXVIII The physical and biological effects of high-frequency sound-waves of great intensity. Lond, Edinb, Dublin Philos Mag J Sci 4 (22):417–436
4. Flynn HG (1964) Physics of acoustic cavitation in liquids. In: Mason WP (ed) Physical acoustics, vol 1, Part B. Academic Press, New York, pp 57–172
5. Rozenberg LD (1971) High-intensity ultrasonic fields. Plenum Press, New York
6. Neppiras EA (1980) Acoustic cavitation. Phys Rep 61(3):159–251
7. Young FR (1989) Cavitation. McGraw-Hill, London
8. Leighton TG (1994) The acoustic bubble. Academic Press, London
9. Brennen EG (1995) Cavitation and bubble dynamics. Oxford University Press, New York
10. Young FR (2005) Sonoluminescence. CRC Press, Boca Raton
11. Mason TJ, Lorimer JP (1988) Sonochemistry. Wiley
12. Mason TJ (ed) (1999) Advances in sonochemistry, vol 5. Jai Press, Stamford
13. Lauterborn W, Kurz T, Mettin R, Ohl C-D (1999) Experimental and theoretical bubble dynamics. Adv Chem Phys 110: 295–380
14. Ohl C-D, Kurz T, Geisler R, Lindau O, Lauterborn W (1999) Bubble dynamics, shock waves and sonoluminescence. Phil Trans R Soc Lond A 357:269–294
15. Mettin R (2007) From a single bubble to bubble structures in acoustic cavitation. In: Kurz T, Parlitz U, Kaatze U (eds) Oscillations, waves and interactions. Universitätsverlag Göttingen, Göttingen, pp 171–198
16. Lauterborn W, Kurz T (2010) Physics of bubble oscillations. Rep Prog Phys 73:106501
17. Lauterborn W, Mettin R (2015) Acoustic cavitation: bubble dynamics in high-power ultrasonic fields. In: Gallego-Juárez JA, Graff KF (eds) Power ultrasonics. Elsevier, pp 37–78
18. Mettin R, Cairós C (2016) Bubble dynamics and observations. In: Ashokkumar M et al (eds) Handbook of ultrasonics and sonochemistry. Springer Science + Business Media, Singapore
19. Harvey EN, McElroy WD, Whiteley AH (1947) On cavity formation in water. J Appl Phys 18(2):162–172
20. Fox FE, Herzfeld KF (1954) Gas bubbles with organic skin as cavitation nuclei. J Acoust Soc Am 26(6):984–989
21. Crum LA (1982) Nucleation and stabilization of microbubbles in liquids. Appl Sci Res 38 (1):101–115
22. Yasui K, Tuziuti T, Kanematsu W, Kato K (2016) Dynamic equilibrium model for a bulk nanobubble and a microbubble partly covered with hydrophobic material. Langmuir 32 (43):11101–11110
23. Keller AP (1974) Investigations concerning scale effects of the inception of cavitation. In: Proceedings I mechanical engineering conference on cavitation, pp 109–117
24. Reuter F, Lesnik S, Ayaz-Bustami K, Brenner G, Mettin R (2018) Bubble size measurements in different acoustic cavitation structures: filaments, clusters, and the acoustically cavitated jet. Ultrason Sonochem. Available online 16 May 2018. https://doi.org/10.1016/j.ultsonch.2018.05.003
25. Minnaert M (1933) On musical air bubbles and the sounds of running water. Phil Mag Ser 7 (16):235–248
26. Parlitz U, Englisch V, Scheffczyk C, Lauterborn W (1990) Bifurcation structure of bubble oscillators. J Acoust Soc Am 88:1061
27. Hilgenfeldt S, Grossmann S, Lohse D (1999) Sonoluminescence light emission. Phys Fluids 11:1318
28. Hilgenfeldt S, Brenner MP, Grossmann S, Lohse D (1998) Analysis of Rayleigh-Plesset dynamics for sonoluminescing bubbles. J Fluid Mech 365:171–204
29. Prosperetti A, Lezzi A (1986) Bubble dynamics in a compressible liquid. Part 1. First-order theory. J Fluid Mech 168:457–478
30. Kamath V, Prosperetti A, Egolfopoulos FN (1993) A theoretical study of sonoluminescence. J Acoust Soc Am 94(1):248–260
31. Yasui K (1997) Alternative model of single-bubble sonoluminescence. Phys Rev E 56:6750

32. Keller JB, Miksis M (1980) Bubble oscillations of large amplitude. J Acoust Soc Am 68:628
33. Mettin R, Cairós C, Troia A (2015) Sonochemistry and bubble dynamics. Ultrason Sonochem 25:24–30
34. Thiemann A, Holsteyns F, Cairos C, Mettin R (2017) Sonoluminescence and dynamics of cavitation bubble populations in sulfuric acid. Ultrason Sonochem 34:663–676
35. Blake FG (1949) Harvard University Acoustic Research Laboratory, Tech. Mem. No. 12, 1949 (unpublished)
36. Noltingk BE, Neppiras EA (1950) Cavitation produced by ultrasonics. Proc Phys Soc Lond, Sect B 63(9):674
37. Louisnard O, Gomez F (2003) Growth by rectified diffusion of strongly acoustically forced gas bubbles in nearly saturated liquids. Phys Rev E 67:036610
38. Lauterborn W, Mettin R (1999) Nonlinear bubble dynamics—response curves and more. In: Crum LA, Mason TJ, Reisse JL, Suslick KS (eds) Sonochemistry and sonoluminescence; Proceedings of the NATO advanced study institute, Leavenworth (WA), USA, 18–29 Aug 1997. Kluwer Academic Publishers, Dordrecht, pp 63–72
39. Franc J-P, Michel J-M (2006) Fundamentals of cavitation. Springer science & Business media, Berlin
40. Van Wijngaarden L (1972) One-dimensional flow of liquids containing small gas bubbles. Ann Rev Fluid Mech 4:369–394
41. Caflisch RE, Miksis MJ, Papanicolaou GC, Ting L (1985) Effective equations for wave propagation in bubbly liquids. J Fluid Mech 153:259–273
42. Commander KW, Prosperetti A (1989) Linear pressure waves in bubbly liquids: comparison between theory and experiments. J Acoust Soc Am 85:732–746
43. Akhatov I, Parlitz U, Lauterborn W (1996) Towards a theory of self-organization phenomena in bubble-liquid mixtures. Phys Rev E 54:4990
44. Louisnard O (2012) A simple model of ultrasound propagation in a cavitating liquid. Part I: theory, nonlinear attenuation and traveling wave generation. Ultrason Sonochem 19:56–65
45. Louisnard O (2012) A simple model of ultrasound propagation in a cavitating liquid. Part II: primary Bjerknes force and bubble structures. Ultrason. Sonochem. 19:66–76
46. Cairós C, Schneider J, Pflieger R, Mettin R (2014) Effects of argon sparging rate, ultrasonic power, and frequency on multibubble sonoluminescence spectra and bubble dynamics in NaCl aqueous solutions. Ultrason Sonochem 21:2044–2051
47. Mettin R, Cairós C (2019) Leuchtende Blasen. Phys Unserer Zeit 50(1):38–42
48. Taylor GI (1950) The instability of liquid surfaces when accelerated in a direction perpendicular to their planes. Proc R Soc Lond A 201:192–196
49. Plesset MS (1954) On the stability of fluid flows with spherical symmetry. J Appl Phys 25 (1):96–98
50. Birkhoff G (1954) Note on Taylor instability. Q Appl Math 12(3):306–309
51. Birkhoff G (1956) Stability of spherical bubbles. Q Appl Math 13(4):451–453
52. Plesset MS, Mitchell TP (1956) On the stability of the spherical shape of a vapor cavity in a liquid. Q Appl Math 13(4):419–430
53. Strube HW (1971) Numerische Untersuchungen zur Stabilität nichtsphärisch schwingender Blasen. Acustica 25:289–303
54. Kornfeld M, Suvorov L (1944) On the destructive action of cavitation. J Appl Phys 15 (6):495–506
55. Hilgenfeldt S, Lohse D, Brenner MP (1996) Phase diagrams for sonoluminescing bubbles. Phys Fluids 8:2808
56. Versluis M, Goertz DE, Palanchon P, Heitman IL, van der Meer SM, Dollet B, de Jong N, Lohse D (2010) Microbubble shape oscillations excited through ultrasonic parametric driving. Phys Rev E 82(2):026321
57. Eller A, Flynn HG (1965) Rectified diffusion during nonlinear pulsations of cavitation bubbles. J Acoust Soc Am 37(3):493–503
58. Fyrillas MM, Szeri AJ (1994) Dissolution or growth of soluble spherical oscillating bubbles. J Fluid Mech 277:381–407

59. Bjerknes VFK (1906) Fields of force. Columbia University Press, New York
60. Matula TJ, Cordry AM, Roy RA, Crum LA (1997) Bjerknes force and bubble levitation under single-bubble sonoluminescence conditions. J Acoust Soc Am 102:1522–1527
61. Akhatov I, Mettin R, Ohl C-D, Parlitz U, Lauterborn W (1997) Bjerknes force threshold for stable single bubble sonoluminescence. Phys Rev E 55:3747–3750
62. Mettin R, Akhatov I, Parlitz U, Ohl CD, Lauterborn W (1997) Bjerknes forces between small cavitation bubbles in a strong acoustic field. Phys Rev E 56:2924–2931
63. Crum LA (1975) Bjerknes forces on bubbles in a stationary sound field. J Acoust Soc Am 57 (6):1363–1370
64. Cairós C, Mettin R (2017) Simultaneous high-speed recording of sonoluminescence and bubble dynamics in multibubble fields. Phys Rev Lett 118(6):064301
65. Levich VG (1962) Physicochemical hydrodynamics. Prentice-Hall, Englewood Cliffs
66. Klyachko LS (1934) Heating and ventilation. USSR J Otopl I Ventil (4)
67. Magnaudet J, Legendre D (1998) The viscous drag force on a spherical bubble with a time-dependent radius. Phys Fluids 10(3):550–554
68. Krefting D, Mettin R, Lauterborn W (2002) Kräfte in akustischen Kavitationsfeldern (Forces in acoustic cavitation fields). In Jekosch U (ed) Fortschritte der Akustik—DAGA 2002, Bochum. DEGA, Oldenburg, pp 260–261
69. Apfel RE (1981) Acoustic cavitation prediction. J Acoust Soc Am 69(6):1624–1633
70. Church CC (1988) Prediction of rectified diffusion during nonlinear bubble pulsations at biomedical frequencies. J Acoust Soc Am 83(6):2210–2217
71. Mettin R (2005) Bubble structures in acoustic cavitation. In: Doinikov AA (ed) Bubble and particle dynamics in acoustic fields: modern trends and applications. Research Signpost, Kerala, pp 1–36
72. Gaitan D, Crum LA, Church CC, Roy RA (1992) Sonoluminescence and bubble dynamics for a single, stable, cavitation bubble. J Acoust Soc Am 91:3166–3183
73. Hiller R, Putterman SJ, Barber BP (1992) Spectrum of synchronous picosecond sonoluminescence. Phys Rev Lett 69:1182
74. Barber BP, Hiller RA, Löfstedt R, Putterman SJ, Weninger KR (1997) Defining the unknowns of sonoluminescence. Phys Rep 281:65–143
75. Crum LA (2015) Resource paper: sonoluminescence. J Acoust Soc Am 138:2181–2205
76. Gompf B, Günther R, Nick G, Pecha R, Eisenmenger W (1997) Resolving sonolumines-cence pulse width with time-correlated single photon counting. Phys Rev Lett 79:1405
77. Chen W, Huang W, Liang Y, Gao X, Cui W (2008) Time-resolved spectra of single-bubble sonoluminescence in sulfuric acid with a streak camera. Phys Rev E 78(3):035301
78. Hiller R, Weninger K, Putterman SJ, Barber BP (1994) Effect of noble gas doping in single-bubble sonoluminescence. Science 266(5183):248–250
79. Schneider J, Pflieger R, Nikitenko SI, Shchukin D, Möhwald H (2010) Line emission of sodium and hydroxyl radicals in single-bubble sonoluminescence. J Phys Chem A 115 (2):136–140
80. Flannigan DJ, Suslick KS (2005) Plasma line emission during single-bubble cavitation. Phys Rev Lett 95:044301
81. Flannigan DJ, Suslick KS (2005) Plasma formation and temperature measurement during single-bubble cavitation. Nature 434(7029):52
82. Lepoint T, Lepoint-Mullie F, Henglein A (1999) Single bubble sonochemistry. In: Crum LA et al (eds) Sonochemistry and sonoluminescence. Kluwer Academic Publishers, Dordrecht, pp 285–290
83. Verraes T, Lepoint-Mullie F, Lepoint T, Longuet-Higgins M (2000) Experimental study of the liquid flow near a single sonoluminescent bubble. J Acoust Soc Am 108:117
84. Troia A, Madonna Ripa D, Lago S, Spagnolo R (2004) Evidence for liquid phase reactions during single bubble acoustic cavitation. Ultrason Sonochem 11:317
85. Didenko YT, Suslick KS (2002) The energy efficiency of formation of photons, radicals and ions during single-bubble cavitation. Nature 418(6896):394

86. Mettin R, Lindinger B, Lauterborn W (2002) Bjerknes-Instabilität levitierter Einzelblasen bei geringem statischen Druck (Bjerknes-instability of levitated single bubbles at low static pressure). In: Jekosch U (ed) Fortschritte der Akustik—DAGA 2002, Bochum. DEGA, Oldenburg, pp 264–265

87. Rosselló JM, Dellavale D, Bonetto FJ (2013) Energy concentration and positional stability of sonoluminescent bubbles in sulfuric acid for different static pressures. Phys Rev E 88:033026

88. Matula TJ, Roy RA, Mourad PD, McNamara WB III, Suslick KS (1995) Comparison of multibubble and single-bubble sonoluminescence spectra. Phys Rev Lett 75:2602

89. Benjamin TB, Ellis AT (1966) The collapse of cavitation bubbles and the pressures thereby produced against solid boundaries. Phil Trans Roy Soc Lond A 260:221–240

90. Calvisi M, Lindau O, Blake JR, Szeri AJ (2007) Shape stability and violent collapse of microbubbles in acoustic traveling waves. Phys Fluids 19:047101

91. Vuong VQ, Szeri AJ, Young DA (1999) Shock formation within sonoluminescence bubbles. Phys Fluids 11:10–17

92. Schanz D, Metten B, Kurz T, Lauterborn W (2012) Molecular dynamics simulations of cavitation bubble collapse and sonoluminescence. New J Phys 14:113019

93. Xu H, Eddingsaas NC, Suslick KS (2009) Spatial separation of cavitating bubble populations: the nanodroplet injection model. J Am Chem Soc 131:6060–6061

94. Xu H, Glumac NG, Suslick KS (2010) Temperature inhomogeneity during multibubble sonoluminescence. Angew. Chemie 122(6):1097–1100

95. Lechner C, Koch M, Lauterborn W, Mettin R (2017) Pressure and tension waves from bubble collapse near a solid boundary: a numerical approach. J Acoust Soc Am 142 (6):3649–3659

96. Blake JR, Hooton MC, Robinson PB, Tong RP (1997) Collapsing cavities, toroidal bubbles and jet impact. Phil Trans Roy Soc Lond A 355:537–550

97. Reuter F, Gonzalez-Avila SR, Mettin R, Ohl C-D (2017) Flow fields and vortex dynamics of bubbles collapsing near a solid boundary. Phys Rev Fluids 2:064202

98. Reuter F, Mettin R (2018) Electrochemical wall shear rate microscopy of collapsing bubbles. Phys Rev Fluids 3:063601

99. Plesset MS, Chapman RB (1971) Collapse of an initially spherical vapour cavity in the neighbourhood of a solid boundary. J Fluid Mech 47(2):283–290

100. Lauterborn W, Bolle H (1975) Experimental investigations of cavitation-bubble collapse in the neighbourhood of a solid boundary. J Fluid Mech 72(2):391–399

101. Philipp A, Lauterborn W (1998) Cavitation erosion by single laser-produced bubbles. J Fluid Mech 361:75–116

102. Krefting D, Mettin R, Lauterborn W (2004) High-speed observation of acoustic cavitation erosion in multibubble systems. Ultrason Sonochem 11:119–123

103. Fuchs FJ (2015) Ultrasonic cleaning and washing of surfaces. In: Gallego-Juárez JA, Graff KF (eds) Power ultrasonics. Elsevier, pp 577–610

104. Mason TJ (2016) Ultrasonic cleaning: an historical perspective. Ultrason Sonochem 29:519–523

105. Reuter F, Mettin R (2016) Mechanisms of single bubble cleaning. Ultrason Sonochem 29:550–562

106. Blake JR, Keen GS, Tong RP, Wilson M (1999) Acoustic cavitation: the fluid dynamics of non–spherical bubbles. Phil Trans R Soc Lond A 357:251

107. Pearson A, Blake JR, Otto SR (2004) Jets in bubbles. J Eng Math 48:391–412

108. Lauterborn W, Lechner C, Koch M, Mettin R (2018) Bubble models and real bubbles: Rayleigh and energy-deposit cases in a Tait-compressible liquid. IMA J. Appl. Math 83 (4):556–589

109. Supponen O, Obreschkow D, Tinguely M, Kobel P, Dorsaz N, Farhat M (2016) Scaling laws for jets of single cavitation bubbles. J Fluid Mech 802:263–293

110. Brujan EA, Noda T, Ishigami A, Ogasawara T, Takahira H (2018) Dynamics of laser-induced cavitation bubbles near two perpendicular rigid walls. J Fluid Mech 841:28–49

111. Ohl SW, Ohl CD (2016) Acoustic cavitation in a microchannel. In: Ashokkumar M et al (eds) Handbook of ultrasonics and sonochemistry. Springer Science + Business Media, Singapore, pp 99–135

112. Koch M, Lechner Ch, Reuter F, Köhler K, Mettin R, Lauterborn W (2016) Numerical modeling of laser generated cavitation bubbles with the finite volume and volume of fluid method, using OpenFOAM. Comput Fluids 126:71–90

113. Lindau O, Lauterborn W (2003) Cinematographic observation of the collapse and rebound of a laser-produced cavitation bubble near a wall. J Fluid Mech 479:327–348

114. Falkovich G (2011) Fluid mechanics, a short course for physicists. Cambridge University Press

115. Blake JR, Leppinen DM, Wang Q (2015) Cavitation and bubble dynamics: the Kelvin impulse and its applications. Interface Focus 5(5):20150017

116. Wang QX, Blake JR (2010) Non-spherical bubble dynamics in a compressible liquid. Part 1. Travelling acoustic wave. J Fluid Mech 659:191–224

117. Nowak T, Mettin R (2014) Unsteady translation and repetitive jetting of acoustic cavitation bubbles. Phys Rev E 90:033016

118. Hatanaka S, Hayashi S, Choi P-K (2010) Sonoluminescence of alkali-metal atoms in sulfuric acid: comparison with that in water. Jpn J Appl Phys 49:07HE01

119. Yasui K (2018) Acoustic cavitation and bubble dynamics. Springer Briefs in Molecular Science—Ultrasound and Sonochemistry, Springer International Publishing

Chapter 2
Sonoluminescence

2.1 Introduction

The previous chapter presented physical characterizations of cavitation bubbles on the microscopic scale, looking, e.g., on the bubble shape, on its stability and evolution, and on the way bubble dynamics can explain energy focusing that leads to sonochemistry and sonoluminescence. These latter two phenomena are macroscopic manifestations of acoustic cavitation and can also serve to characterize bubbles and their activity. Micro- and macroscopic measurements are fully complementary in the study of cavitation: While the former will bring information on each type of bubble separately, in a time-resolved manner, the latter will deliver an overall spatially averaged picture of the effects of the whole bubble population.

The present chapter focuses on the light emitted by cavitation bubbles at collapse, the so-called sonoluminescence (SL), and in particular on its measurements and on the information that can be derived from them. The various theories developed to explain SL emission will not be detailed here. The reader is referred to the book Sonoluminescence [1] by Young which also provides an excellent review of the knowledge on SL up to 2005.

A first part presents experimental evidence that allowed to affirm that a plasma forms in cavitation bubbles at collapse. The various information (temperatures, density, and electron density) that can be derived from SL spectra using tools from plasma spectroscopy are then described. A third part deals with the way measurements of SL intensity under pulsed ultrasound can be used to estimate some bubble size distribution.

Focus will be put on aqueous solutions and multibubble sonoluminescence since these systems are the most relevant for sonochemical applications.

2.2 Experimental Evidence of the Formation of a Plasma in Cavitation Bubbles

It is now widely recognized, both by sonochemists and by the plasma community [2], that a plasma forms in cavitation bubbles at collapse. First irrefutable experimental evidence of it was the observation in single-bubble sonoluminescence (SBSL) spectra of sulfuric acid in the presence of 67 mbar of a rare gas of the emission from electronically excited ions (Ar^+, Kr^+, Xe^+, and O_2^+) [3, 4] and from high-energy excited rare gas atoms (Fig. 2.1). The energy of the emitting excited states is too high for their population to have a thermal origin. For instance, an energy of >13 eV (translating into temperatures >150,000 K) is necessary to populate the observed emitting states of Ar atoms. These emissions can only be explained by excitation via high-energy particle (e.g., electrons) collisions originating from a hot plasma core [3]. Further evidence of the plasma formation was given by O_2^+ emission. Indeed, the energy required to dissociate O_2 molecule (5.1 eV) is lower than its ionization energy (12.1 eV), which means that in a thermal process, O_2 molecules would dissociate, and O_2^+ would not be formed.

Though often considered, the plasma origin of the MBSL emission from aqueous solutions was not that straightforward since no emission from excited ions that would be formed in the plasma was observed. It is to be noted that this absence does not mean that, e.g., Ar^{+*} or H_2O^{+*} ions do not form but more probably that they are non-radiatively de-excited ("quenched") by collisions with water molecules that are abundant in cavitation bubbles due to the high vapor pressure of water.

It was recently that definite evidence of a plasma formation during multibubble cavitation in water saturated with a rare gas was provided by the observation of emission from highly excited OH ($C^2\Sigma^+$) radicals (Fig. 2.2) [5]. The emission of OH ($C^2\Sigma^+$–$A^2\Sigma^+$) cannot be accounted for in a thermal model, since it would

Fig. 2.1 SBSL spectrum from 85% H_2SO_4 with 67 mbar and an acoustic pressure of 1.7 bar. Reprinted from [4] with permission; © American Physical Society

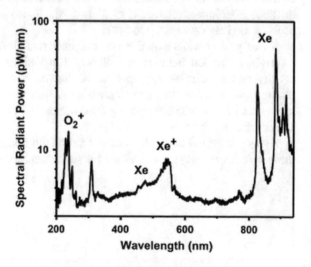

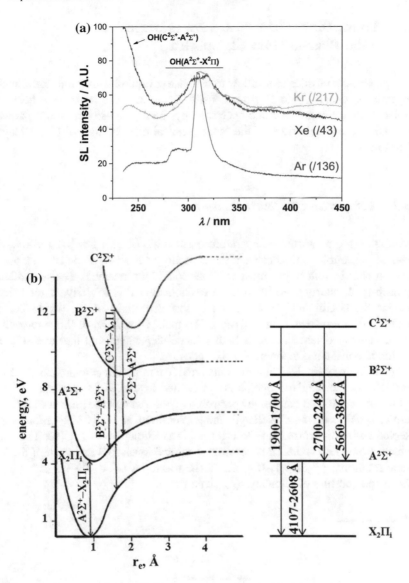

Fig. 2.2 a SL spectra of water under different rare gases at 20 kHz (adapted from [5]; © Wiley);
b simplified energy diagram of OH radical and wavelength range of its emissions

require 16.1 eV to excite water molecules at the right level. On the contrary, this
emission has been reported in spectra of electrical discharges [6, 7] through water
vapor and OH ($C^2\Sigma^+$) formation can be readily explained by electron impacts. The
emission of OH ($C^2\Sigma^+$–$A^2\Sigma^+$) was reported [5] to be very strong at high US
frequency, and to increase when the gas was changed from Ar to Kr to Xe, cor-
relating with the decrease in the ionization potential of the gas.

2.3 Plasma Diagnostics to Derive Information on the Plasma from SL Spectra

Usual plasma diagnostics tools allow to estimate several plasma parameters, such as temperatures, density, and electron density. Some of these tools have been used coupled with sonoluminescence spectroscopy, though mostly in some particular cases due to the peculiarities of the "sonochemical plasma" that will be detailed in the following.

2.3.1 Temperature Determination

Determining the maximum temperature reached at collapse has been a long-time quest. Since sonoluminescence emission occurs at the last stage of collapse, its spectrum should reflect the most extreme conditions reached. The first adopted approach (particularly used in SBSL measurements) was to derive a temperature from the SL continuum shape, assuming that the origin of SL would be, e.g., blackbody or Bremsstrahlung emission. The main drawbacks of this approach are that the calculated temperature is highly model-dependent and that one sole contribution is considered to shape the SL spectrum.

On the contrary, molecular emissions provide a more direct approach since their shape reflects the relative populations of excited levels. The following paragraph does not intend to be a course on spectroscopy but to recall the most useful facts to interpret spectra. In the wavelength range measured in SL (UV to near-IR), the observed emissions correspond to electronic transitions $(e' \rightarrow e'')$ (see Fig. 2.2b). Each electronic level comprises several vibrational levels (v') and each of the latter several rotational (J') levels (Fig. 2.3). The transition $(e', v', J' \rightarrow e'', v'', J'')$ produces a spectral line of intensity $I(\lambda)$ given by:

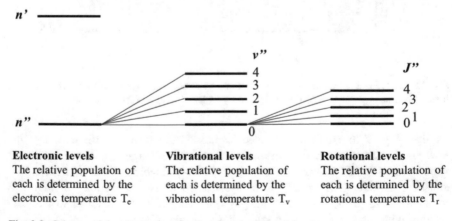

Electronic levels
The relative population of each is determined by the electronic temperature T_e

Vibrational levels
The relative population of each is determined by the vibrational temperature T_v

Rotational levels
The relative population of each is determined by the rotational temperature T_r

Fig. 2.3 Scheme of the electronic, vibrational, and rotational levels

$$I(\lambda) = n_{e'v'J'}A_{e'v'J'-e''v''J''}h\nu_0\Phi(\lambda - \lambda_0) \tag{2.1}$$

where $n_{e'v'J'}$ is the population of the emitting rovibronic level, $A_{e'v'J'-e''v''J''}$ is the transition probability between the two levels (tabulated), $h\nu_0$ is the energy difference between the two levels and $\Phi(\lambda - \lambda_0)$ the line broadening function. It is generally assumed that the electronic, vibrational, and rotational levels all follow a Boltzmann law, each at the corresponding temperature (electronic temperature T_e, vibrational T_v, and rotational T_r). The population of the emitting rovibronic level is then given by:

$$n_{e'v'J'} = \frac{g_{e'}e^{-\frac{E_{e'}}{kT_e}}}{Q_e}\frac{g_{v'}e^{-\frac{E_{v'}}{kT_v}}}{Q_v}\frac{g_{J'}e^{-\frac{E_{J'}}{kT_r}}}{Q_r}n_{tot} \tag{2.2}$$

where n_{tot} is the total population of the considered species, k is Boltzmann constant, $g_{e'}, g_{v'}$ and $g_{J'}$ are the (tabulated) degeneracies of the states, Q_e is the electronic partition function defined as (vibrational, Q_v, and rotational, Q_r, partition functions are defined similarly):

$$Q_e = \sum_e g_e e^{-\frac{E_e}{kT_e}} \tag{2.3}$$

Thus, the population of each emitting level and, consequently, the intensity of each spectral line and the global shape of the transition are determined by the three temperatures T_e, T_v, and T_r. The general case in a non-equilibrium plasma is $T_e > T_v > T_r \geq T_{gas}$ where T_{gas} is the gas temperature, i.e., the translation temperature. If time scales are long enough, thermal equilibrium can be reached and all temperatures have the same value.

Spectroscopic codes exist that calculate the emission shape for a given species and given temperatures, like for instance Lifbase [8] or Specair [9]. They also take into account broadening of the emissions by several parameters including the experimental broadening and the pressure broadening.

Such an approach has been used on C_2 Swan ($d^3\Pi_g - a^3\Pi_u$) bands in the pioneering works of Flint and Suslick [10] in silicone oil and of Didenko et al. [11] in an aqueous 10^{-3} M benzene solution, both sonicated at 20 kHz under Ar. Both assumed unicity of the temperature, but considering the way it was determined (it was chosen to tally with the relative intensities of C_2 bands) it corresponds to a vibrational temperature. Fitting of the spectra resulted in the temperatures 5075 ± 156 K for silicone oil and 4300 ± 200 K for the aqueous benzene solution. The same approach was applied some years later on the emissions of Fe, Cr, and Mo in the MBSL 20 kHz spectra of metal carbonyls in silicone oil [12] under Ar, leading to similar temperatures: $4700-5100 \pm 400$ K.

Concentrated acids were also studied because of their very bright sonoluminescence that allows high-resolution spectroscopy. MBSL spectra of concentrated phosphoric acid sonicated at 20 kHz under He [13] show the molecular emissions of OH $(A^2\Sigma^+-X^2\Pi)$ and PO $(B^2\Sigma^+-X^2\Pi)$, the fitting of which gave temperatures of 9500 and 4000 K, respectively. This difference in temperature was explained by the emission arising from symmetrically collapsing bubbles for OH and by fast moving bubbles for PO. In concentrated sulfuric acid sonicated at 20 kHz under Ar, the emission of excited Ar atoms (4p-4 s manifold) indicates a temperature of 8000 K [14].

Interestingly, OH $(A^2\Sigma^+-X^2\Pi)$ emission spectrum in MBSL [13] in concentrated phosphoric acid is quite similar to its emission in SBSL spectra of a rapidly moving single bubble [15] in 65% H_3PO_4 regassed with 67 mbar Ar (see Fig. 2.4) and temperatures derived from simulations are very close. The moving SBSL study allows to investigate the effect of the acoustic pressure: As expected, its increase leads to a net increase in OH temperature. As for a change in the rare gas nature, going over the series from He to Xe leads to an increase in temperature from 6000 to >10,000 K. Xu and Suslick [15] attributed this effect to the strong decrease in thermal conductivity of the gas limiting the heat losses during bubble collapse. Another explanation would be the strong decrease in the gas ionization potential that makes plasma formation easier and increases its electron energy.

In all these studies, the unicity of temperature was assumed in the fitting procedure, and in most cases, this approach lead to consistent results. More recently, however, a study [16] of the influence of the ultrasonic frequency on SL spectra of aqueous *tert*-butanol solutions under Ar showed that fitting with a single temperature was not possible in this case and that $T_v > T_r$. For instance, temperatures of $T_v = 6300 \pm 1000$ K and $T_r = 4800 \pm 1000$ K were obtained at 20 kHz under Ar in the concentration range 0.05–0.4 M, where C_2 Swan bands were highest. At high frequency (204, 362, and 613 kHz), higher vibrational temperatures were obtained [16], indicating that the plasma had a higher electron temperature. For example, the simulation of Swan bands lead to $T_v = 8000 \pm 1000$ K and $T_r = 4000 \pm 1000$ K

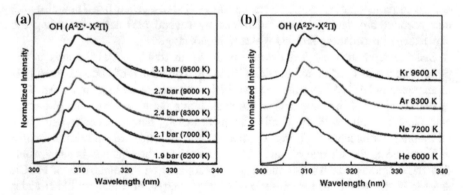

Fig. 2.4 a Emission of OH $(A^2\Sigma^+-X^2\Pi)$ in SBSL spectra of a rapidly moving bubble driven at different acoustic pressures in 65% H_3PO_4 regassed with 67 mbar Ar (reproduced with permission from [15]; © American Physical Society); **b** emission of OH $(A^2\Sigma^+-X^2\Pi)$ in SBSL spectra of a rapidly moving bubble driven at 2.4 bar in 65% H_3PO_4 regassed with 67 mbar of a rare gas (reproduced with permission from [15]; © American Physical Society)

at 362 kHz under Ar in the *tert*-butanol concentration range 1.10^{-3}–5.10^{-3} M. Besides, replacing Ar by Xe at 20 kHz leads to a very high T_v (14,000 K). This effect was explained by the relatively low ionization potential of Xe providing a higher electron temperature in the non-equilibrium plasma generated at bubble collapse. Another interesting finding was that an increase in *tert*-butanol concentration induced a decrease in T_v and in the intensity of the SL continuum, up to a certain concentration range where it then stayed constant (and where C_2 emission has its maximum intensity). Obviously, an increase in concentration of this volatile solute leads to an increase in the number of *tert*-butanol molecules inside cavitation bubbles, molecules that cushion the bubble collapse, leading to a lower energy concentration, and consume this energy. Hence, the lower SL intensity and T_v. Much higher concentrations of *tert*-butanol were needed at low US frequency to quench SL, which was attributed to the smaller bubble size (i.e., to the higher bubble surface/volume ratio) at high ultrasonic frequency (see Figs. 1.3 and 1.11), favorable to vaporization of volatile molecules into bubbles, and to the fact that high-frequency bubbles remain active for many more cycles than 20 kHz ones, thus accumulating more hydrocarbon decomposition products.

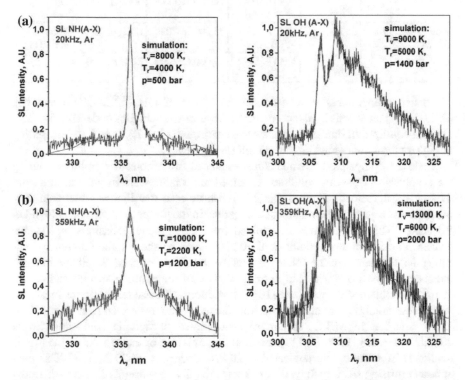

Fig. 2.5 Molecular emission of OH ($A^2\Sigma^+$–$X^2\Pi$) and NH ($A^3\Pi$–$X^3\Sigma^-$) in SL spectra of 0.1 M ammonia solutions sonicated under Ar at 20 and 359 kHz and their simulations with Specair [9]. Reproduced with permission from [17]; © Royal Society of Chemistry

Table 2.1 Rovibronic temperatures estimated from the simulations of molecular emissions in SL spectra from aqueous solutions

Solution	Gas	US frequency, kHz	Transition	T_v, K	T_r, K	References
Benzene 10^{-3} M	Ar	20	C$_2$ Swan (d$^3\Pi_g$–a$^3\Pi_u$)	4300 ± 200		Didenko et al. [11]
tert-butanol 0.05–0.4 M	Ar	20	C$_2$ Swan (d$^3\Pi_g$–a$^3\Pi_u$)	6300 ± 1000	4800 ± 1000	Pflieger et al. [16]
tert-butanol 1.10^{-3} – 5.10^{-3} M	Ar	204/362/ 613	C$_2$ Swan (d$^3\Pi_g$–a$^3\Pi_u$)	5800/8000/ 5000 (± 1000)	5800/4000/ 4000 (± 1000)	Pflieger et al. [16]
tert-butanol 0.12 M	Xe	20	C$_2$ Swan (d$^3\Pi_g$–a$^3\Pi_u$)	14,000 ± 1000	2500 ± 1000	Pflieger et al. [16]
Ammonia 0.1 M	Ar	20	OH (A$^2\Sigma^+$– X$^2\Pi$)	9000 ± 1000	5000 ± 500	Pflieger et al. [17]
			NH (A$^3\Pi$– X$^3\Sigma^-$)	7000 ± 1000	4000 ± 500	
Ammonia 0.1 M	Ar	359	OH (A$^2\Sigma^+$– X$^2\Pi$)	13,000 ± 2000	6000 ± 1000	Pflieger et al. [17]
			NH (A$^3\Pi$– X$^3\Sigma^-$)	10,000 ± 1000	2200 ± 500	

Similar temperatures were estimated by fitting of OH (A$^2\Sigma^+$–X$^2\Pi$) and NH (A$^3\Pi$–X$^3\Sigma^-$) emissions in SL spectra of 0.1 M ammonia solutions under Ar (Fig. 2.5). The non-equilibrium state of the plasma was confirmed ($T_v > T_r$), as well as the reaching of higher vibrational temperatures at high US frequency.

Table 2.1 compares the temperatures estimated from molecular emissions during the sonolysis of aqueous solutions. Obtained temperatures are in general in a relatively good agreement. From Table 2.1, it can be seen that higher vibrational temperatures (more extreme conditions) are reached in the presence of Xe and at high US frequency. The temperatures determined from Swan band emissions in aqueous benzene and tert-butanol solutions at 20 kHz under Ar show some difference, indicating that the nature of the solute impacts the conditions reached at collapse but also the chemical reactions that lead to the formation of the emitting species and consequently the obtained excited state [16]. Interestingly, all these temperatures measured for aqueous solutions are not very far from the temperatures derived from OH (A$^2\Sigma^+$–X$^2\Pi$) emission in MBSL spectra of concentrated phosphoric acid sonicated at 20 kHz under He (9500 K) [13], or from the emission of excited Ar atoms (4p-4s manifold) in sulfuric acid sonicated at 20 kHz under Ar (8000 K) [14]. This may appear surprising since such media are known to offer very bright SL due to their low volatility and the high solubility of sonolysis products that make bubble collapse particularly efficient [18]. This apparent inconsistency may be due to the formation of an optically opaque plasma core, whereby measured emissions would occur from the outer "cooler" part of the bubbles.

Spectroscopy belongs to the very few experimental methods to estimate temperatures. Another one is a chemical determination, where a temperature is estimated by comparing the yields of two reactions, the reaction rates of which are known as a function of temperature. For instance, Ciawi et al. [19, 20] studied the kinetics of recombination of methyl radicals in the sonolysis of aqueous *tert*-butanol solutions under Ar at different US frequencies. The chemical temperatures they obtained were 3400 ± 200 K for 20 kHz, 4300 ± 200 K for 355 kHz, and 3700 ± 200 K for 1056 kHz. Obviously, chemical temperatures are significantly lower than vibrational temperatures and therefore than electron temperatures. This difference can be explained by the fact that the chemical temperature is a mean temperature that is not spatially and temporally well defined, whereas SL occurs at the hottest point in the bubble life. What is more, this chemical model assumes thermal equilibrium, and parallel reaction pathways are neglected. Despite everything, it is noteworthy that the same trend of higher temperatures at high frequency is observed by both methods.

2.3.2 Current Limitations of the Fitting of SL Emissions

Molecular emissions in SL present severe drawbacks when it comes to fitting them, especially in aqueous solutions. First of all, they lie on top of an intense continuum. Second, they are dim and very broad. Despite these limitations, in many cases it appears possible to fit them, as seen above. Yet, one can find several examples in the literature where fitting was not feasible. For instance, in Fig. 2.5 simulations of OH $(A^2\Sigma^+-X^2\Pi)$ and NH $(A^3\Pi-X^3\Sigma^-)$ come very close to the experimental spectra at 20 kHz but do not manage to reproduce them properly at 359 kHz. This impossibility to fit properly these emissions at high frequency (and at any frequency in the presence of Xe) was attributed by Ndiaye et al. [21] to a deviation from Boltzmann law of the populations of excited vibrational levels of OH $(A^2\Sigma^+)$, whereby higher excited levels would be overpopulated. Flannigan and Suslick [22] also reported a non-Boltzmann distribution, namely of the vibrational levels of SO, in SBSL spectra from 80 wt% aqueous sulfuric acid solutions containing dissolved neon, while it was Boltzmann for a 65 wt% sulfuric acid solution.

Another factor was recently [17] underlined that would need further developments to enable huge improvement in the fitting of SL spectra: the taking into consideration of Stark effects in spectroscopy softwares. Indeed, in a dense plasma like the one formed in cavitation bubbles, the main broadening mechanism is the collisional one that consists in collisions with neutral species (the pressure broadening) but also with charged species (Stark effects). The perturbation of the energy levels by the electric field of charged species leads to two effects: a broadening and a shift in emission wavelengths. Unfortunately, these Stark effects are not easy to quantify for molecular species and the corresponding equations are pretty much species dependent [23]. These effects being negligible in low and atmospheric pressure plasmas, they are at the moment not taken into account in existing spectroscopy softwares. Devoted developments would be needed to enable SL fitting especially at high frequency and in the presence of Xe [17]. Indeed, in these cases, the higher ionization degree of the intrabubble plasma means stronger Stark effects and much more broadened emissions. Interestingly, the strong broadening of emissions was not only observed in MBSL in aqueous solutions, but also in the spectra of a moving single-bubble in phosphoric acid [15], in the presence of Kr and Xe (i.e., the rare gases with lower ionization potentials), rendering the spectral fitting impossible.

2.3.3 Pressure/Density Determination

The following section is adapted from the discussion on the pressure determination in [17]. It is generally reported that a pressure of several hundred bars is reached during acoustic cavitation, though direct measurements of it are lacking. The usual approach [1] leading to a rough estimate of the maximum reached pressure is to consider adiabatic compression of a bubble filled with a pure ideal gas. This approach strongly relies on the value of the maximum temperature and on the assumption of the unicity of this temperature—which, as seen above, is generally not supported by SL spectra, especially in aqueous solutions. To take an example, considering a pure Ar bubble in the adiabatic compression model, a peak pressure of 146 atm is estimated for a maximum temperature of 2200 K, 1130 atm for 5000 K, and 6350 atm for 10,000 K. This simple approach can thus serve to roughly estimate the conditions at collapse, keeping in mind the underlying hypotheses and the problematic of the maximum temperature.

Spectral analysis (measurements of line widths or shifts) was used to experimentally determine the maximum pressure. Indeed, an increase in pressure, i.e., in the intrabubble density, leads to disturbance of the emitting species, of its lifetime and of the energies of its emitting levels, effects that result in line asymmetry, shift and broadening [24]. Quantification of these effects allows to estimate the relative gas density inside the bubble at the time of emission. Since the value that speaks to most people is the pressure, the obtained gas density is then often converted into a pressure by assuming a certain temperature value.

In the first method, the pressure is derived from line shifts. Lepoint-Mullie et al. [25] measured Rb resonance line shifts during the 20-kHz sonication of RbCl solutions (aqueous solutions and in 1-octanol) under Ar or Kr and compared these shifts, in the range 0.4–0.7 nm, with tabulated ones as a function of relative density. A relative density of 18 ± 2 was deduced therefrom. The same approach was used by McNamara et al. [12] on chromium carbonyl solutions in silicon oil sonicated at 20 kHz under Ar. Comparison of the experimental shifts of the excited Cr* emission with reference shifts from ballistic compression data gave a relative density of 19 ± 2, showing a perfect agreement with the value of Lepoint-Mullie et al. [25] From this relative density and taking into account a temperature of 4700 K, they estimated a pressure of approximately 300 bar.

The second used approach tries to correlate the line broadening to the relative gas density. The underlying assumption is that the main sources of line broadening are instrumental broadening and collisional broadening and that the others can be neglected. Sodium showing a strong emission in SL, it was naturally most often taken as a probe. Sehgal et al. [26] calculated a relative density of 36–50 in aqueous alkali metal salt solutions sonicated at 460 kHz under Ar. In a similar system (NaCl solutions sonicated at 138 kHz under Ar), Choi et al. [27] estimated the maximum relative density around 59.5. Using perfect gas law and 4300 K for the maximum temperature, they translated this density into a pressure of 873 atm. In these two works, the pressure effect on the emission line broadening was calculated in the model of collisional broadening. A similar but more empirical approach was used at lower US frequencies (22 and 44 kHz) by Kazachek and Gordeychuk [28] who studied the SL of NaCl aqueous solutions under Ar and estimated pressures of 800–1200 bar, thus similar. Also, the pressure values derived from a spectral fit with Specair [9] software of the emissions of NH ($A^3\Pi$–$X^3\Sigma^-$) and OH ($A^2\Sigma^+$–$X^2\Pi$) in SL spectra of aqueous ammonia solutions under Ar [17] were of the same order of magnitude: The pressure estimated from NH ($A^3\Pi$–$X^3\Sigma^-$) emission was 500 bar at 20 kHz and 1200 bar at 359 kHz, that estimated from OH ($A^2\Sigma^+$–$X^2\Pi$) emission 1400 bar at 20 kHz and 2000 bar at 359 kHz. All these values of relative gas density estimated from line broadening under the assumption that the latter would result from a combination of instrumental and collisional broadenings are clearly much higher than those derived from line shifts. This apparent discrepancy confirms the occurrence of strong Stark effects and that their contribution to line broadening cannot be neglected. Stark broadening is highly dependent on the nature of the emitting species [23], which explains the very different "pressure" values derived from the emissions of OH and NH in [17].

Interestingly, in one SBSL study, [29] in sulfuric acid partially regassed with Ar, the Stark effects were taken into account. In this work, three contributions to Ar line broadening were considered: the instrumental one, the pressure broadening, and the Stark broadening. The three contributions were decoupled and the reached density was derived and converted into a pressure of 1400 bar. Though conditions achieved in a single bubble in sulfuric acid may be far from those reached in MBSL bubbles in aqueous solutions, this study exemplifies the possibility to separate the various broadening contributions to derive information on the plasma, here in the relatively simple Ar case.

2.3.4 Electron Density and Electron Temperature

In more conventional plasmas, electron properties are usually measured by an electrostatic probe such as a Langmuir probe [30]. Due to the peculiarities of the sonochemical plasma (in particular, its very small size), this technique is not available to sonochemists. Other techniques involve emission spectroscopy and some of them may be used with SL spectra. In usual plasmas, the electron temperature has been obtained using emission intensities from two different electronic levels of a rare gas [31–33], though the latter was contained in a small quantity in the plasma, contrary to usual SL cases. The spectral line broadening caused by the Stark effect is also the basis of a very efficient plasma diagnostics method that can be extrapolated to the sonochemical plasma. Indeed, for any emitter, the broadening and shift in wavelength will depend on electron temperature and electron density [23, 34]. This dependency is complex and species dependent, but the method is promising.

Measurement-based estimations of the electron density and temperature in SL are very scarce. In water, it was long hindered by the lack of molecular emissions in the spectra. First attempts thus relied on the light continuum: Lepoint et al. [35] estimated electronic temperature ($T_e \approx 20{,}000$ K) and density ($N_e = 10^{25}$ m^{-3}) from the SBSL continuum of an Ar bubble in water, considering its origin as being radiative recombination and bremsstrahlung. This very first estimation was based on hypotheses that condition the obtained values, namely that the bubble size at the moment of emission would be 2 μm and the light pulse duration 50 ps. The same group also estimated the electron temperature T_e from MBSL spectra of water saturated with Ar, by comparing the intensities of a line to that of the adjacent continuum [36]. They observed an increase in this temperature with the US frequency, but unfortunately data were not published and only a conference abstract can be found.

In 2010, Flannigan and Suslick [37] analyzed the emission lines from electronically excited Ar atoms in SBSL spectra of concentrated sulfuric acid containing Ar at 5% of saturation, and determined the plasma electron density as a function of the acoustic pressure from the shape of the emission and in particular from its deviation from a Lorentzian shape (Fig. 2.6). They also derived a

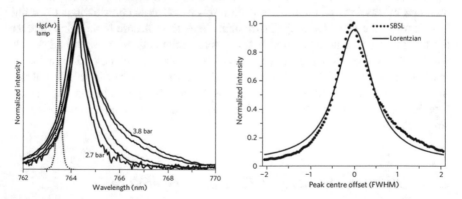

Fig. 2.6 *Left*: Ar emission line profiles as a function of the acoustic driving pressure, p_a, in SBSL spectra of concentrated sulfuric acid containing Ar at 5% of saturation; *Right*: comparison of one SBSL Ar line profile with a Lorentzian curve. Figures reprinted with permission from [37]; © Springer Nature

Table 2.2 Temperatures and electron densities derived from Ar emission in SBSL spectra of concentrated sulfuric acid containing Ar at 5% of saturation [37], as a function of the applied acoustic pressure

p_a, bar	T_{Ar}, K	N_e, cm^{-3}
2.7	7000	4.10^{17}
3.0	10,000	1.10^{18}
3.3	13,000	2.10^{19}
3.6	15,000	5.10^{20}
3.8	16,000	4.10^{21}

temperature from the fitting of Ar emission. In this particular case, fitting with one single temperature was possible. Table 2.2 presents the values they obtained as a function of the applied acoustic pressure (p_a): both temperature and electron density strongly increase with p_a.

Apart from the works by Lepoint et al. [35, 36], the determination of electron density and electron temperature in MBSL in aqueous solutions remains for the moment an almost blank territory. Due to the limitations discussed above, classical plasma diagnostics methods could not be applied until now. Nevertheless, several hints can be found in the literature that the electron energy can be quite high in cavitation bubbles in water, especially at high US frequency. For instance, it was shown that a significant number of O_2 molecules can be dissociated in cavitation bubbles during water sonolysis at high frequency under Ar [38]. The dissociation energy of O_2 being 5.2 eV, this means that a significant number of electrons have an energy ≥ 5.2 eV. The same phenomenon was observed for N_2 molecules: Water sonolysis in the presence of Ar and N_2 leads to NH ($A^3\Pi-X^3\Sigma^-$) emission in MBSL spectra at high frequency but not at 20 kHz [39], whereby the most probable formation mechanism of NH requires N_2 dissociation and an energy of 9.8 eV. The same gas nature effect was observed in this study as in [5, 15, 16, 21], namely that changing Ar for Xe increases the electron energy: While under Ar–N_2 NH ($A^3\Pi$–$X^3\Sigma^-$) emission was not observed at 20 kHz, it was clearly present when sonication was performed under Xe–N_2 mixture.

2.4 Emission of Non-volatile Solutes in SL

It has been known since 1970 that sonication can excite non-volatile solutes and that corresponding light emission can be observed on SL spectra [40]. Several studies thus focused on SL spectra of salt solutions to try to derive information either on the conditions reached in the cavitation bubbles at collapse, or on the mechanism of formation of the excited species that emit light. They mostly focused on two ion families: alkali metal ions and lanthanide ions.

2.4.1 Alkali Metals

In air-saturated aqueous solutions of salts [41], the SL spectra consist of a featureless continuum. On the contrary, in the presence of a rare gas, emissions arising from electronic transitions appear in the SL spectra: OH $(A^2\Sigma^+ - X^2\Pi)$ that is typical of aqueous solutions and emission from excited metal (e.g., Na, K, and Rb) atoms accompanied by a blue satellite (Fig. 2.7). The latter corresponds to the transition of an alkali metal–Ar exciplex, a van der Waals molecule formed within the cavitation bubbles [25].

The mechanism of Na^* (or other electronically excited alkali metals) formation was long a subject of debate, whether inside the bubbles or at their interface. The nowadays accepted mechanism is summarized in Fig. 2.8: Some volume of solution is mechanically added to a collapsing bubble (by droplet injection); therein, the salt molecules are released in the plasma phase and homolytically cleaved, producing Na and Cl. The metal atoms are then electronically excited by three-body reactions, leading to Na^*. As for the electronically excited Na–Ar* exciplex, it would be formed following a three-body collision with two rare gas atoms. It is to be noted that two different excited states of the exciplex are populated, leading to the blue satellite observed around 557 nm (B–X transition) and to the apparent line distortion of Na^* peak toward higher wavelengths (A–X transition).

Emission spectra of electronically excited alkali metals were used to derive information on the conditions reached at collapse, since their emission occurs inside cavitation bubbles. In particular, several attempts aimed at estimating the intrabubble density, with the two methods presented above. The first one relies upon the shift in wavelength of emission: Using it, Lepoint-Mullie et al. [25] determined the intrabubble density in aqueous solutions sonicated at 20 kHz under Ar to be 18 ± 2. The second method is based on the assumption that the pressure would be

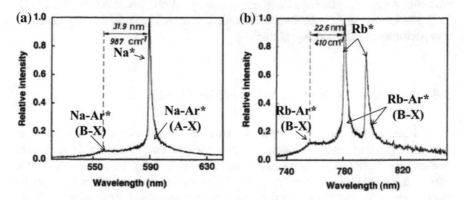

Fig. 2.7 SL spectra of NaCl (**a**) and RbCl (**b**) solutions sonicated at 20 kHz under Ar. Adapted with permission from [25]; © Elsevier

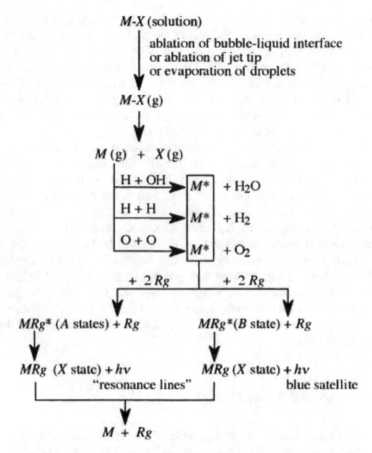

Fig. 2.8 Scheme of the reaction pathways leading to Na* and Na–Ar* formations as proposed by Lepoint-Mullie et al. [25]. Reprinted with permission from Lepoint-Mullie et al. [25]; © Elsevier

the main source of peak broadening and led to much higher relative densities, of the order of 36–80 [26–28]. As discussed above, the latter values are overestimated because this method neglects the strong broadening due to the presence of charged species.

Recent spectroscopic studies of alkali metal solutions revealed that the global emission of the metal and of its exciplex is in fact composed by the superposition of two components that do not emit at the same time [42]: a thin, not-shifted emission of the excited alkali metal atom and a broadened red-shifted emission of the exciplex. Not only is the time of emission different, but also the spatial distribution, suggesting that both emissions would arise from different bubble populations. Interestingly, the broadened component was found to be prominent under He and Ne, while the narrow one was enhanced under Ar, Kr, or Xe. Further research is needed to clarify this observation.

It was also shown in recent literature that different bubble populations were responsible for the emissions of Na* and of the SL continuum. Abe and Choi [43] reported a difference in the timing distributions of both SL emissions during the sonication at 137 kHz of a NaCl 2M aqueous solution saturated with Ar. Spectroscopy of SL indicated that the evolution of sonoluminescing bubble population and of Na-emitting bubbles with gas flow, power, and frequency were different [44]. Moreover, visual observations of SL from both concentrated sulfuric acid solutions [45, 46], phosphoric acid [47], and in aqueous solutions [43, 48] evidenced a spatial separation of continuum emission and alkali metal emission (blue continuum and orange Na* emission). The former was attributed to higher-temperature bubbles and the latter to lower-temperature bubbles [43]. It was also suggested based on a bubble radius simulation that continuum emission would result from smaller bubbles than Na-atom emission. The latter assumption was confirmed by high-speed imaging in sulfuric acid at 23 kHz under Xe: [47] Different cavitation bubble populations were observed in the zones of different SL colors, namely slow and spherically collapsing bubbles in the blue-emitting zones, fast bubbles subject to liquid jetting during their collapse in the red-emitting zone. However, while it is clear that Na* emission arises only from jetting bubbles, it has also been reported in Xe saturated phosphoric acid that SL continuum emission can also be observed from large bubbles having a translational motion and therefore showing jetting [49].

2.4.2   Lanthanides and Uranyl

A second chemical family has been a topic of interest in SL studies, namely the trivalent lanthanide ions, and the similarly behaving uranyl ion (UO_2^{2+}). Indeed, these ions can be excited by two different mechanisms: either by photon absorption or by collisions with highly energetic species (e.g., in radiolysis), with excitation yields that depend on the ion nature. The emission spectrum is the same in both cases. In a solution submitted to ultrasonic irradiation, the two mechanisms of excitation are a priori possible.

The feasibility of the excitation of lanthanide ions under ultrasound was shown by Sharipov et al. [50] for single-bubble and multibubble configurations at 20 kHz. Their work was confirmed and extended to high frequencies by Pflieger et al. [51]. In SBSL, the sole excitation mechanism is by absorption of the photons of sonoluminescence. In MBSL, photoexcitation is prominent for Ce^{3+}, while collisions are more efficient to excite Tb^{3+} and Eu^{3+}. Sharipov et al. explained this difference by collisional excitation occurring inside bubbles and made possible by droplet injections. As for uranyl ions, they are mainly excited by photons at low uranyl concentrations ($\leq 10^{-2}$ M) while the collisional mechanism becomes important above 0.03 M [52].

A major difference between sonoexcitation and classical photoexcitation was brought to light [51, 52] namely the presence of an extensive quenching of excited species. This non-radiative de-excitation was attributed to the enhanced collisional (Stern–Volmer) quenching in the overheated zone around cavitation bubbles, by chemical reactions with sonolytical products and by non-radiative relaxation via coupling with OH vibrations of the solvent. This quenching leads to a decrease of the luminescence apparent yield. Its extent can be reduced by complexation of the lanthanide ions by citrate ions (respectively of uranyl by phosphate ions).

The ultrasonic frequency was shown to play an important role on the observed yield of emission [52]. In the more simple case of Ce^{3+} (where excitation occurs by photon absorption with a yield of 1), for instance, the SL yield, defined as the ratio of the number of emitted photons to that of absorbed photons, was 0.9 at 204 kHz versus 0.4 at 20 kHz, indicating a much more intense quenching at low frequency, possibly due to the larger size of the bubbles leading to higher quenching. On the contrary, in the case of terbium, whose excitation mechanism is mainly by collisions, the apparent SL yield decreased with the US frequency, suggesting a decrease in the excitation efficiency.

2.5   Bubble Size Estimation by SL Intensity Measurement Under Pulsed Ultrasound

The emission of SL by cavitation bubbles at collapse was also taken advantage of to estimate bubble sizes and bubble size distributions, in a method developed at the University of Melbourne [53]. This method is based on the dissolution of bubbles in a pulsed acoustic field and measurement of the SL light intensity. The principle is depicted in Fig. 2.9.

During the pulse on time (t_{on}), bubbles form and grow. Its value is chosen to allow the formation of a population of active bubbles while avoiding non-desired interactions between them. It is followed by a pulse off time, t_{off}, during which bubbles dissolve. Some bubbles can also coalesce, which would impact the bubble size distribution, but this effect is generally neglected [53–56]. The size of the

Fig. 2.9 Evolution of the bubble population under pulsed ultrasound. Reprinted with permission from [53]; © American Chemical Society

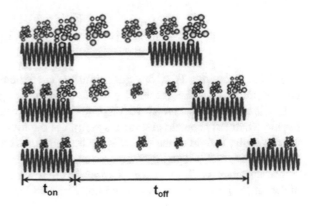

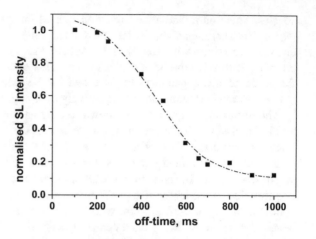

Fig. 2.10 Typical SL intensity versus pulse off-time curve; 0.5 M NaCl solution, Ar bubbling, 355 kHz. Reprinted with permission from [56]; © American Chemical Society

bubbles at the end of t_{on} determines whether they fully dissolve during t_{off} or only partly, turning into bubble nuclei for the subsequent t_{on}. If t_{off} is very short, the dissolution underwent by bubbles (whatever their size) is only partial and they all remain as nuclei for the next t_{on}. As t_{off} is increased, more and more bubbles dissolve below a critical size range. Thus, the number of nuclei present at the beginning of t_{on} decreases, and so does the number of active bubbles, which illustrates in a decrease of the SL (or SCL) intensity. Figure 2.10 presents typical SL versus "off-time" data observed for a 0.5 M NaCl solution under Ar at 355 kHz. Around 700 ms, the curve shows a second inflexion point, after which the decrease in SL intensity almost stops: An off time of 700 ms is sufficient to decrease the size of the maximum number of active cavitation bubbles below a critical size range allowing them to act as nuclei in the subsequent on-pulse. In the range of t_{off} between the two inflexion points, each increment in t_{off} leads to the dissolution of bubbles with a corresponding size.

The bubble size and bubble size distribution are then derived from the SL intensity evolution with t_{off}, using the dissolution equation of a single stationary bubble [57]:

$$\left(\frac{DC_s}{\rho_g R_0^2}\right)t = \frac{1}{3}\left(\frac{RT\rho_g R_0}{2M\gamma} + 1\right) \tag{2.4}$$

In this equation, D is the gas diffusion coefficient, C_s is the dissolved gas concentration, ρ_g is the gas density inside the bubble, R_0 is the radius of the bubble before it starts to dissolve (it corresponds to ambient radius, i.e., the radius of the bubble when the acoustic pressure is zero), t is the total dissolution time (t_{off}), M is the molecular weight of the gas, R is the universal gas constant, T is the temperature of the liquid, and γ is the surface tension.

Measurements of bubble size distributions in water and 1.5 mM SDS at 515 kHz under air [53] pointed out that the addition of SDS lead to a decrease in bubble size and in the width of the bubble size distribution. It was also shown [55], in air-saturated solutions sonicated at 575 kHz, that sonochemically (SC) active bubbles (measured in luminol solutions) were smaller than sonoluminescence (SL) bubbles (in water). Another important result of these studies on bubble size and bubble size distribution was the experimental confirmation that the radius of SC bubbles decreased with an increase in ultrasonic frequency, accompanied by a narrowing of the size distribution (Fig. 2.11) [55]. It is remarkable that despite the assumptions made in this technique of bubble size determination (in particular, the funding one, that nothing else but dissolution happens during the pulse off time), the obtained sizes, at least in this particular case, are in pretty good agreement with the calculated ones in Fig. 1.11.

Other experimental parameters that do affect the bubble size are the dissolved gas nature and the presence of salts in solution. Measurements at 515 kHz [54] indicated an increase in the bubble size when the gas was changed from helium to air to argon. The same study showed that an increase in the salt (NaCl, KCl, and NaNO$_3$) concentration led to smaller bubbles. Brotchie et al. [54] explained this effect arguing that the presence of salts in solution decreased the gas solubility. These salts were also shown to reduce the extent of bubble coalescence, confirming the importance of this mechanism in the bubble growth.

The effect of dissolved gas was further investigated [56] in aqueous NaCl solutions sonicated at 355 kHz and submitted to continuous Ar or He gas flow (whereas previous studies considered pre-saturated solutions). Similarly to previous

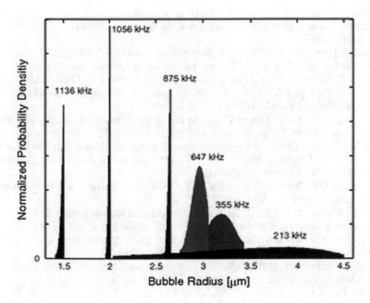

Fig. 2.11 Bubble size distributions for 213, 355, 647, 875, 1056, and 1136 kHz (the data for 875, 1056, and 1136 kHz have been scaled down by a factor of 4) in an air-saturated luminol solution. Reprinted with permission from [55]; © American Physical Society

results on Ar or He pre-saturated solutions, the bubble size was reported to decrease with increasing NaCl concentration. The continuous gas flow strongly enhanced this decrease. Thus, it is the combination of several parameters that determines the bubble size, including the gas concentration, controlled by the salt concentration, and the number of cavitation nuclei, that are introduced, e.g., by a continuous gas bubbling. Besides, the gas diffusion coefficient also appears to play a role in defining the bubble size: The combination of high gas solubility and high gas diffusion coefficient allows a faster bubble growth in each expansion cycle and subsequent bigger bubble sizes.

References

1. Young FR (2005) Sonoluminescence. CRC Press, New York
2. Bruggeman PJ, Kushner MJ, Locke BR, Gardeniers JGE, Graham WG, Graves DB, Hofman-Caris R, Maric D, Reid JP, Ceriani E, Rivas DF, Foster JE, Garrick SC, Gorbanev Y, Hamaguchi S, Iza F, Jablonowski H, Klimova E, Kolb J, Krcma F, Lukes P, Machala Z, Marinov I, Mariotti D, Thagard SM, Minakata D, Neyts EC, Pawlat J, Petrovic ZL, Pflieger R, Reuter S, Schram DC, Schroter S, Shiraiwa M, Tarabova B, Tsai PA, Verlet JRR, von Woedtke T, Wilson KR, Yasui K, Zvereva G (2016) Plasma-liquid interactions: a review and roadmap. Plasma Sources Sci Technol 25
3. Flannigan DJ, Suslick KS (2005) Plasma formation and temperature measurement during single-bubble cavitation. Nature 434:52–55
4. Flannigan DJ, Suslick KS (2005) Plasma line emission during single-bubble cavitation. Phys Rev Lett 95
5. Pflieger R, Brau HP, Nikitenko SI (2010) Sonoluminescence from OH($C^2\Sigma^+$) and OH($A^2\Sigma^+$) radicals in water: evidence for plasma formation during multibubble cavitation. Chem Eur J 16:11801–11803
6. Michel A (1957) Das $C^2\Sigma^+$- $A^2\Sigma^+$- Bandensystem von OH, Zeitschrift Für Naturforschung Part A-Astrophysik Physik Und Physikalische Chemie 12:887–896
7. Carlone C, Dalby FW (1969) Spectrum of hydroxyl radical. Can J Phys 47:1945–1957
8. Luque J, Crosley DR (1999) LIFBASE: database and spectral simulation. In: SRI international report MP 99-009
9. Laux CO, Spence TG, Kruger CH, Zare RN (2003) Optical diagnostics of atmospheric pressure air plasmas. Plasma Sources Sci Technol 12:125–138
10. Flint EB, Suslick KS (1991) The temperature of cavitation. Science 253:1397–1399
11. Didenko YT, McNamara WB, Suslick KS (1999) Hot spot conditions during cavitation in water. J Am Chem Soc 121:5817–5818
12. McNamara WB, Didenko YT, Suslick KS (1999) Sonoluminescence temperatures during multi-bubble cavitation. Nature 401:772–775
13. Xu HX, Glumac NG, Suslick KS (2010) Temperature inhomogeneity during multibubble sonoluminescence. Angewandte Chemie-International Edition 49:1079–1082
14. Eddingsaas NC, Suslick KS (2007) Evidence for a plasma core during multibubble sonoluminescence in sulfuric acid. J Am Chem Soc 129:3838–3839
15. Xu HS, Suslick KS (2010) Molecular emission and temperature measurements from single-bubble sonoluminescence. Phys Rev Lett 104:244301
16. Pflieger R, Ndiaye AA, Chave T, Nikitenko SI (2015) Influence of ultrasonic frequency on Swan band sonoluminescence and sonochemical activity in aqueous tert-butyl alcohol solutions. J Phys Chem B 119:284–290

17. Pflieger R, Ouerhani T, Belmonte T, Nikitenko SI (2017) Use of NH ($A^3\Pi_i$-$X^3\Sigma^-$) sonoluminescence for diagnostics of nonequilibrium plasma produced by multibubble cavitation. Phys Chem Chem Phys 19:26272–26279

18. Suslick KS, Eddingsaas NC, Flannigan DJ, Hopkins SD, Xu HX (2011) Extreme conditions during multibubble cavitation: sonoluminescence as a spectroscopic probe. Ultrason Sonochem 18:842–846

19. Ciawi E, Ashokkumar M, Grieser F (2006) Limitations of the methyl radical recombination method for acoustic cavitation bubble temperature measurements in aqueous solutions. J Phys Chem B 110:9779–9781

20. Ciawi E, Rae J, Ashokkumar M, Grieser F (2006) Determination of temperatures within acoustically generated bubbles in aqueous solutions at different ultrasound frequencies. J Phys Chem B 110:13656–13660

21. Ndiaye AA, Pflieger R, Siboulet B, Molina J, Dufreche JF, Nikitenko SI (2012) Nonequilibrium vibrational excitation of OH radicals generated during multibubble cavitation in water. J Phys Chem A 116:4860–4867

22. Flannigan DJ, Suslick KS (2013) Non-Boltzmann population distributions during single-bubble sonoluminescence. J Phys Chem B 117:15886–15893

23. Gigosos MA (2014) Stark broadening models for plasma diagnostics. J Phys D Appl Phys 47

24. Margenau H, Lewis M (1959) Structure of spectral lines from plasmas. Rev Mod Phys 31:569–615

25. Lepoint-Mullie F, Voglet N, Lepoint T, Avni R (2001) Evidence for the emission of 'alkali-metal-noble-gas' van der Waals molecules from cavitation bubbles. Ultrason Sonochem 8:151–158

26. Sehgal C, Steer RP, Sutherland RG, Verrall RE (1979) Sonoluminescence of argon saturated alkali-metal salt-solutions as a probe of acoustic cavitation. J Chem Phys 70:2242–2248

27. Choi PK, Abe S, Hayashi Y (2008) Sonoluminescence of Na atom from NaCl solutions doped with ethanol. J Phys Chem B 112:918–922

28. Kazachek MV, Gordeychuk TV (2009) Estimation of the cavitation peak pressure using the Na D-line structure in the sonoluminescence spectra. Tech Phys Lett 35:193–196

29. Flannigan DJ, Hopkins SD, Camara CG, Putterman SJ, Suslick KS (2006) Measurement of pressure and density inside a single sonoluminescing bubble. Phys Rev Lett 96

30. Derkaoui N, Rond C, Gries T, Henrion G, Gicquel A (2014) Determining electron temperature and electron density in moderate pressure H_2/CH_4 microwave plasma. J Phys D Appl Phys 47

31. Gicquel A, Chenevier M, Hassouni K, Tserepi A, Dubus M (1998) Validation of actinometry for estimating relative hydrogen atom densities and electron energy evolution in plasma assisted diamond deposition reactors. J Appl Phys 83:7504–7521

32. Shatas AA, Hu YZ, Irene EA (1992) Langmuir probe and optical-emission studies of Ar, O_2, and N_2 plasmas produced by an electron-cyclotron resonance microwave source. J Vac Sci Technol A-Vac Surfaces Films 10:3119–3124

33. Mehdi T, Legrand PB, Dauchot JP, Wautelet M, Hecq M (1993) Optical-emission diagnostics of an RF magnetron sputtering discharge. Spectrochim Acta Part B-Atomic Spectrosc 48:1023–1033

34. Belmonte T, Noel C, Gries T, Martin J, Henrion G (2015) Theoretical background of optical emission spectroscopy for analysis of atmospheric pressure plasmas. Plasma Sources Sci Technol 24

35. Lepoint-Mullie F, De Pauw D, Lepoint T, Supiot P, Avni R (1996) Nature of the "extreme conditions" in single sonoluminescing bubbles (vol 100, p 12140). J Phys Chem A 103 (1999):3346

36. Lepoint T, Lepoint-Mullie F, Avni R (1996) Plasma diagnostics and sonoluminescence. J Acoust Soc Am 100:2677

37. Flannigan DJ, Suslick KS (2010) Inertially confined plasma in an imploding bubble. Nat Phys 6:598–601

38. Pflieger R, Chave T, Vite G, Jouve L, Nikitenko SI (2015) Effect of operational conditions on sonoluminescence and kinetics of H_2O_2 formation during the sonolysis of water in the presence of Ar/O_2 gas mixture. Ultrason Sonochem 26:169–175
39. Ouerhani T, Pflieger R, Ben Messaoud E, Nikitenko SI (2015) Spectroscopy of sonoluminescence and sonochemistry in water saturated with N_2 – Ar mixtures. J Phys Chem B 119:15885–15891
40. Taylor KJ, Jarman PD (1970) Spectra of sonoluminescence. Aust J Phys Aust J Phys 23:319
41. Wall M, Ashokkumar M, Tronson R, Grieser F (1999) Multibubble sonoluminescence in aqueous salt solutions. Ultrason Sonochem 6:7–14
42. Nakajima R, Hayashi Y, Choi PK (2015) Mechanism of two types of Na emission observed in sonoluminescence. Jpn J Appl Phys 54
43. Abe S, Choi PK (2009) Spatiotemporal separation of Na-atom emission from continuum emission in sonoluminescence. Jpn J Appl Phys 48
44. Cairos C, Schneider J, Pflieger R, Mettin R (2014) Effects of argon sparging rate, ultrasonic power, and frequency on multibubble sonoluminescence spectra and bubble dynamics in NaCl aqueous solutions. Ultrason Sonochem 21:2044–2051
45. Hatanaka S, Hayashi S, Choi PK (2010) Sonoluminescence of Alkali-Metal atoms in sulfuric acid: comparison with that in water. Jpn J Applied Phys 49
46. Xu HX, Eddingsaas NC, Suslick KS (2009) Spatial separation of cavitating bubble populations: the nanodroplet injection model. J Am Chem Soc 131:6060−+
47. Thiemann A, Holsteyns F, Cairos C, Mettin R (2017) Sonoluminescence and dynamics of cavitation bubble populations in sulfuric acid. Ultrason Sonochem 34:663–676
48. Sunartio D, Yasui K, Tuziuti T, Kozuka T, Iida Y, Ashokkumar M, Grieser F (2007) Correlation between Na* emission and "chemically active" acoustic cavitation bubbles. Chem Phys Chem 8:2331–2335
49. Cairos C, Mettin R (2017) Simultaneous high-speed recording of sonoluminescence and bubble dynamics in multibubble fields. Phys Rev Lett 118
50. Sharipov GL, Gainetdinov RK, Abdrakhmanov AM (2003) Sonoluminescence of aqueous solutions of lanthanide salts. Russ Chem Bull 52:1969–1973
51. Pflieger R, Schneider J, Siboulet B, Mohwald H, Nikitenko SI (2013) Luminescence of trivalent lanthanide ions excited by single-bubble and multibubble cavitations. J Phys Chem B 117:2979–2984
52. Pflieger R, Cousin V, Barre N, Moisy P, Nikitenko SI (2012) Sonoluminescence of Uranyl ions in aqueous solutions. Chem Eur J 18:410–414
53. Lee J, Ashokkumar M, Kentish S, Grieser F (2005) Determination of the size distribution of sonoluminescence bubbles in a pulsed acoustic field. J Am Chem Soc 127:16810–16811
54. Brotchie A, Statham T, Zhou MF, Dharmarathne L, Grieser F, Ashokkumar M (2010) Acoustic bubble sizes, coalescence, and sonochemical activity in aqueous electrolyte solutions saturated with different gases. Langmuir 26:12690–12695
55. Brotchie A, Grieser F, Ashokkumar M (2009) Effect of power and frequency on bubble-size distributions in acoustic cavitation. Phys Rev Lett 102
56. Pflieger R, Lee J, Nikitenko SI, Ashokkumar M (2015) Influence of He and Ar flow rates and NaCl concentration on the size distribution of bubbles generated by power ultrasound. J Phys Chem B 119:12682–12688
57. Epstein PS, Plesset MS (1950) On the stability of gas bubbles in liquid-gas solutions. J Chem Phys 18:1505–1509

Chapter 3
Sonochemistry

3.1 Introduction

The chemical effects of ultrasound were reported for the first time by Richards and Loomis in 1927 for the processes of dimethyl sulfate hydrolysis and the reduction of potassium iodate by sulfurous acid known as iodine "clock" reaction [1]. However, the most famous sonochemical reaction of water molecule splitting was discovered two years later by Schmitt et al. [2]. This reaction has attracted a lot of attention of researches for several reasons. First, hydroxyl radicals and hydrogen peroxide formed during water sonolysis are widely used for sonochemical oxidation of organic pollutants in aqueous solutions [3]. In addition, hydrogen produced simultaneously with hydrogen peroxide was found to be effective for sonochemical reduction of noble metal ions resulting in highly monodispersed metal nanoparticles without addition of any side reagents [4–6]. Finally, the formation of hydroxyl radicals and hydrogen peroxide is often deployed as a chemical dosimeter to measure the specific acoustic power absorbed by solution submitted to power ultrasound [7, 8]. In fact, the dissociation of water molecule is a strongly endothermic process ($\Delta G = 113$ kcal mol^{-1}). Therefore, appearance of sonolytical products during ultrasonic treatment of water clearly indicates transient cavitation which provides drastic conditions inside the imploding bubble required for the rupture of O–H bond. Spectroscopic studies of multibubble sonoluminescence described in the previous chapter revealed that sonochemical splitting of water molecule occurs via electronic excitation mechanism. Water molecule can be excited to A^1B_1, B^1A_1, and C^1B_1 states [9]. The A^1B_1 and B^1A_1 states are repulsive, and excited $H_2O(A^1B_1)$ and $H_2O(B^1A_1)$ molecules dissociate yielding OH $(X_2\Pi)$ and OH$(A^2\Sigma^+)$ radicals, respectively. By contrast, $H_2O(C^1B_1)$ molecules relax through A^1B_1 or B^1A_1 states. At high ultrasonic frequency and in the presence of easily ionized noble gases, such as Kr and Xe, highly energetic OH$(C^2\Sigma^+)$ state is also formed due to the electron impact in non-equilibrium plasma produced during bubble collapse at these conditions [10]. Hydroxyl radicals in the ground

© The Author(s), under exclusive licence to Springer Nature Switzerland AG 2019   61
R. Pflieger et al., *Characterization of Cavitation Bubbles and Sonoluminescence*,
Ultrasound and Sonochemistry, https://doi.org/10.1007/978-3-030-11717-7_3

state as well as in the excited states enable various sonochemical reactions inside
the bubble and at the bubble/solution interface. Therefore, kinetics of OH• radicals
or H_2O_2 molecules' formation can be used for quantification of acoustic power
delivered to the system. This chapter will focus on the influence of several fundamental parameters, such as ultrasonic frequency, saturating gas, and some soluble
nitrogen compounds on chemical reactivity of multibubble cavitation in homogeneous aqueous media in connection with the recent data on multibubble sonoluminescence overviewed in Chap. 2.

3.2   Sonochemical Dosimetry

Sonochemical water splitting was observed for the first time due to the oxidation of
iodide ion in sonicated aqueous solutions [2]. Since that time, iodometric method is
often used to quantify the sonochemical activity [7, 11]. In general, it is suggested
that iodide ion is oxidized by hydrogen peroxide or by hydroxyl radicals formed
during sonochemical splitting of water molecule:

$$H_2O—))) \rightarrow H + OH \tag{3.1}$$

$$2\,H \rightarrow H_2 \tag{3.2}$$

$$2\,OH \rightarrow H_2O_2 \tag{3.3}$$

$$H_2O_2 + 2\,I^- + 2\,H^+ \rightarrow I_2 + 2H_2O \tag{3.4}$$

$$OH + I^- \rightarrow OH^- + I \tag{3.5}$$

$$I + I^- \rightarrow I_2^- \tag{3.6}$$

$$2\,I_2^- \rightarrow I_2 + 2I^- \tag{3.7}$$

$$I_2 + I^- \rightarrow I_3^- \tag{3.8}$$

This reaction scheme is true for water saturated with noble gases. However,
iodometric method is hardly applicable to aerated aqueous solutions submitted to
power ultrasound. Sonolysis of water in the presence of nitrogen yields nitrous acid
[12] which readily oxidizes iodide ion [13]. Moreover, this reaction is catalyzed by
dissolved oxygen [14]. Consequently, the quantification of kinetic data for this
system becomes uncertain. Obviously, the Fricke sonochemical dosimeter based on
Fe(II) oxidation [7] suffers from a similar drawback.

In terephthalic acid (TA) dosimetry, TA solution reacts specifically with a
hydroxyl radical yielding 2-hydroxyterephthalic acid which can be detected using
fluorescence spectroscopy [15]. It is noteworthy that only part of OH• radicals
produced by cavitation bubble and reaching bubble interface reacts with TA,

whereas a significant fraction gives H_2O_2 after OH• radical recombination. The kinetics of OH• radical formation during multibubble cavitation in water also can be measured using salicylic acid (SA) as radical trapping reagent [16, 17]. The products of SA sonochemical oxidation with OH• radicals, such as 2,3-dihydroxybenzoic acid and 2,5-dihydroxybenzoic acid, can be measured by HPLC technique.

Both TA and SA dosimetric systems are highly sensitive to OH• radicals. However, they require quite complex analytical equipment. Therefore, for many practical uses the simple spectrophotometric method of H_2O_2 concentration measurement with Ti(IV) was used as sonochemical dosimeter [18, 19]. In acidic medium, Ti(IV) ions form stable yellow-colored peroxide complexes allowing quantitative analysis of hydrogen peroxide formed during water sonolysis. It should be mentioned that in the absence of OH• or H_2O_2 scavengers, the rate of H_2O_2 sonochemical formation follows zero-order kinetic law in a wide range of experimental conditions which is convenient for chemical dosimetry [18].

Whatever chemical dosimetric system, the amount of formed sonochemical products should be normalized to specific acoustic power (P_{ac}) in order to compare different sonochemical conditions. Usually, P_{ac} is measured by thermal probe method presuming that most of acoustic power delivered to solution is transformed to heat: [20]

$$P_{ac} = \rho \, C_p \frac{\Delta T}{\Delta \tau} \tag{3.9}$$

where ρ (g mL^{-1}) is the density of the sonicated liquid, C_p (J g^{-1} K^{-1}) is the heat capacity of the liquid, and $\frac{\Delta T}{\Delta \tau}$ (K s^{-1}) is the initial heating rate of sonicated liquid measured under quasi-adiabatic conditions when T increases linearly with time of ultrasonic treatment. Using kinetic data and P_{ac}, it is possible to calculate the yield of the sonochemical reactions (G, μmol kJ^{-1}) useful for comparison of sonochemical efficiency at different experimental conditions.

3.3 Effect of Ultrasonic Frequency

Since a long time, it is recognized that the increase of ultrasonic frequency leads to more efficient OH• radical production during water sonolysis [21]. However, the quantification of frequency effect is still challenging because of the complexity of phenomena occurring in solution when the ultrasonic frequency is increased: diminishing of the resonance bubble size, acceleration of the bubble implosion, modification of bubble size distribution as well as the geometry of the bubble cloud, and increase of the rovibronic temperatures of intrabubble plasma [8, 22–24]. In terms of sonochemical activity, the most suitable approach to compare different frequencies appears to be the use of sonochemical yield, G, defined in Sect. 3.2. Several studies revealed that the yield of hydrogen peroxide during sonolysis of argon-saturated water reached its maximum between 200 and 400 kHz and then progressively diminished [19, 22, 25]. The optimal value of ultrasonic frequency can potentially be related to a significant decrease of the bubble volume with ultrasonic frequency. For example, the linear resonance radius of the cavitation bubble in water is equal to 10.0 μm at 358 kHz and to only 3.3 μm at 1071 kHz [20]. Taken these values as the average bubble size, the bubble volume at 1071 kHz would be almost 30 times smaller than that at 358 kHz, leading to a much smaller relative active volume producing primary products of sonolysis. From Chap. 1, we have seen, however, that the bubbles tend to shift their sizes toward the Blake threshold (depending on the applied driving pressure), which would result in somehow smaller values than the linear resonance radii. This view is corroborated by the bubble size measurements reported at the end of Chap. 2 where SL in pulsed fields had been used. Thus, one might employ the same argument, but for the measured radii (Fig. 2.11) of about 3.2 μm at 358 kHz and 1.5–2 μm at 1071 kHz. Still, the volume change would amount a factor of up to 10. Of course, the absolute active bubble number is assumed similar to this comparison, which might not be precise. In conclusion, the sonochemical activity at different frequencies derives from the superposition of several, partly counteracting phenomena and is difficult to predict. A smooth variation of yield with the frequency, including some optimum, is to be expected if the comparisons are performed in a "fair" fashion, i.e., with comparable powers. However, this problem touches the notoriously difficult issue of scaling acoustic cavitation and its effects, and in many cases experimental parameter tests cannot be fully avoided for optimization.

3.4 Effect of Oxygen

Several research groups reported a significant increase of H_2O_2 sonochemical yield in the presence of oxygen [19, 26, 27]. Maximal yield of H_2O_2 was observed at 20–30 vol.% of oxygen in O_2/Ar gas mixtures whatever the ultrasonic frequency. Similar to neat argon, the optimal frequency for H_2O_2 formation in O_2/Ar gas mixture is at 200–400 kHz as it can be seen from Fig. 3.1. Interesting is that high-frequency ultrasound is approximately 5 times more efficient for sonochemical production of hydrogen peroxide compared to 20 kHz ultrasound.

Enhanced sonochemical efficiency of high-frequency ultrasound is related to intrabubble conditions. Study of multibubble sonoluminescence spectra revealed more efficient O_2 dissociation at 362 kHz compared to 20 kHz which should lead to the increase of $G(H_2O_2)$ [19]. In general, sonochemical formation of H_2O_2 in the presence of oxygen can be described by the following scheme:

$$H_2O—))) \rightarrow H + OH \qquad (3.10)$$

Fig. 3.1 Effect of saturating gas and ultrasonic frequency on the sonochemical yield of hydrogen peroxide. $T = 20\,°C$, gas bubbling at 80 mL min^{-1}, no mechanical stirring. ▨ 20 kHz, $P_{ac} = 33$ W; ▧ 204 kHz, $P_{ac} = 41$ W; ■ 362 kHz, $P_{ac} = 43$ W; ▤ 613 kHz, $P_{ac} = 43$ W. Reproduced from [19]

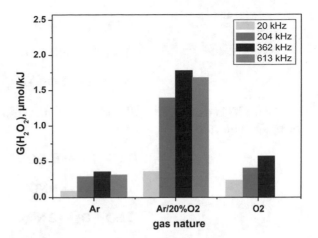

$$O_2 -))) \rightarrow 2\,O \tag{3.11}$$

$$O + H_2O \rightarrow 2\,OH \tag{3.12}$$

$$H + O \rightarrow OH \tag{3.13}$$

$$2\,OH \rightarrow H_2O_2 \tag{3.14}$$

$$H + O_2 \rightarrow HO_2 \tag{3.15}$$

$$2\,HO_2 \rightarrow H_2O_2 + O_2 \tag{3.16}$$

Scavenging of H atoms by molecular oxygen (reaction 3.15) is confirmed by sharp decrease of H_2 formation rate in Ar/20%O_2 compared to pure Ar [19]. Relatively low G(H$_2$O$_2$) in pure O_2 most probably is related to reported argon plasma quenching by O_2 molecules [28].

3.5 Effect of Nitrogen and Ammonia

The sonochemistry of nitrogen in aqueous solutions was pioneered in 1936 by Shultes and Gohr [29]. They reported the formation of HNO$_2$ and NO$_3^-$ ions under the effect of 900 kHz ultrasound in water sparged with air. Later, Mišik and Riesz [30] suggested that H$_2$O$_2$ and NO$_2^-$ were the primary products of water sonolysis in the presence of air and that NO$_3^-$ ion resulted from the secondary oxidation of nitrous acid by hydrogen peroxide. According to Wakeford et al. [31], the highly reactive oxygen required for nitrite formation from molecular nitrogen would come from the dissociation of oxygen molecules inside the cavitation bubble:

$$O_2 \text{—}))) \rightarrow 2\,O \tag{3.17}$$

$$N_2 + O \rightarrow NO + N \tag{3.18}$$

$$N + OH \rightarrow NO + H \tag{3.19}$$

Then, NO is oxidized by OH• radicals originated from water molecule splitting or by O_2 molecules:

$$H_2O \text{—}))) \rightarrow H + OH \tag{3.20}$$

$$NO + OH \rightarrow HNO_2 \tag{3.21}$$

$$2\,NO + O_2 \rightarrow 2\,NO_2 \tag{3.22}$$

$$NO_2 + OH \rightarrow HNO_3 \tag{3.23}$$

In the presence of N_2/Ar gas mixtures, sonochemical processes in water involve the formation of intermediate NH radicals [12] are originated from the intrabubble dissociation of N_2 molecules [32]. H_2, H_2O_2, HNO_2, and NO_3^- species were identified as stable products of sonolysis in studied system [12]. It was shown that the dissociation of N_2 molecules is more efficient at higher ultrasonic frequency in agreement with more drastic intrabubble conditions at high frequency revealed by multibubble sonoluminescence spectroscopy (see Chap. 2).

Formation of NH• radicals was also observed during sonolysis of ammonia solutions saturated with noble gases [33]. Ammonia is known to be volatile, and for that reason its splitting most likely occurred inside the cavitation bubble simultaneously with water molecules:

$$NH_3 \text{—}))) \rightarrow NH + H_2 \tag{3.24}$$

Chemical analysis of ammonia solutions submitted to ultrasound revealed the formation of hydrazine. This process was attributed to the mutual recombination of NH_2• radicals or to NH• radicals scavenging with NH_3 molecules probably at the cavitation bubble/solution interface:

$$H_2O \text{—}))) \rightarrow H + OH \tag{3.25}$$

$$NH_3 + OH \rightarrow NH_2 + H_2O \tag{3.26}$$

$$NH_2 + NH_2 + M \rightarrow N_2H_4 + M \tag{3.27}$$

$$NH + NH_3 + M \rightarrow N_2H_4 + M \tag{3.28}$$

$$NH + OH \rightarrow NO + H_2 \tag{3.29}$$

Similar to other sonochemical systems, yield of hydrazine is larger at high-frequency ultrasound compared to 20 kHz. However, monitoring of hydrogen peroxide indicates that, in contrast to pure water, H_2O_2 does not accumulate during the sonolysis of ammonia solutions, most probably due to the OH• radicals scavenging by NH• radicals (reaction 3.29) and/or NH_3 molecules (reaction 3.26) or because of the rapid reaction of H_2O_2 with hydrazine:

$$N_2H_4 + 2\,H_2O_2 \rightarrow N_2 + 4\,H_2O \tag{3.30}$$

The principal gaseous sonolytical product of ammonia solutions is hydrogen gas. Higher yield of H_2 in ammonia solutions compared to pure water is undoubtedly related to the sonochemical splitting of NH_3 molecules and secondary reactions with nitrogen-containing intermediates.

3.6   Effect of CO and CO_2

Sonochemistry of CO and CO_2 in aqueous media is much less studied than that of O_2 and N_2. Nikitenko et al. [34] reported that the sonication of water with 20 kHz ultrasound in the presence of CO/Ar gas mixture causes a drastic decrease in the H_2O_2 formation rate relative to that in pure Ar. Furthermore, sonolysis at these conditions causes water acidification which is not observed in pure argon. More recently, it was shown that the yield of H_2 during water sonolysis increases drastically in the presence of CO (10 vol.% CO/Ar) [6] indicating that the suppression of H_2O_2 and water acidification is related to OH• radicals scavenging rather than to the decrease of global sonochemical activity:

$$CO + OH \rightarrow CO_2 + H \tag{3.31}$$

$$2\,H \rightarrow H_2 \tag{3.32}$$

$$CO_2 + H_2O \rightleftharpoons H^+ + HCO_3^- \,(pK = 3.60) \tag{3.33}$$

Interesting is that at higher concentration of CO (20 vol.% CO/Ar) sonolysis leads to the formation of carbonaceous products which pointed out the CO disproportionation during bubble collapse [34]. It is worth noting that the amounts of carbonaceous products formed during sonolysis are very sensitive to the experimental conditions. Decreasing the CO concentration to 10% or heating the sonicated water to 45 °C resulted in a sharp decrease in their yield. Chemical analysis revealed that the composition of these products is close to hydrated poly(carbon suboxide). The anhydrous cyclic polymer with a basic formula $(C_3O_2)_n$ is formed in strongly non-equilibrium plasma [35]. At such conditions, endothermic CO disproportioning ($\Delta H = 5.5$ eVmol^{-1}, $E_a = 6$ eVmol^{-1}) can be significantly accelerated by the vibrational excitation of CO molecules:

$$CO(v_1) + CO(v_2) \rightarrow CO_2 + C \qquad (3.34)$$

$$C + CO(v_0) + Ar \rightarrow CCO + Ar \qquad (3.35)$$

$$CCO + CO(v_0) + Ar \rightarrow C_3O_2 + Ar \qquad (3.36)$$

$$n\,C_3O_2 \rightarrow (C_3O_2)_n \qquad (3.37)$$

where v_0 and v_1; v_2 are the ground state and the excited vibrational levels of CO respectively. The similarity of the solid products for the sonochemical and plasma chemical reactions allowed to assume that the mechanism of ultrasonically driven disproportionation of CO is similar to that in non-equilibrium plasma. This hypothesis was confirmed by inverse kinetic isotope effect observed during sonochemical CO disproportionation. In fact, carbonaceous product was found to be enriched with the heavy ^{13}C isotope ($\alpha = {}^{13}C/{}^{12}C = 1.053 - 1.055$). According to semiclassical kinetic isotope effect theory, the reaction products should be enriched by light isotopes because of their higher zero vibrational-level energy [36]. By contrast, the vibrational excitation of CO molecules in non-equilibrium plasma leads to reverse isotope effect since heavy isotopes have higher vibrational temperature. The finding of reverse kinetic isotope effect was the first evidence for non-equilibrium plasma formation during multibubble cavitation. Further studies of sonoluminescence spectroscopy considered in Chap. 2 confirmed this striking phenomenon.

It should also be noted that CO can be formed during sonochemical degradation of organic compounds, such as formic acid [25]. Carbon monoxide molecules enable to diffuse inside the bubble and influence the overall sonochemical mechanism. Thus, C_2 Swan band emission in the sonoluminescence spectra of HCOOH aqueous solutions observed at high-frequency ultrasound was attributed to CO disproportionation followed by the formation of excited C_2^* radical: [37]

$$2\,CO \rightarrow CO_2 + C \qquad (3.38)$$

$$C + C + M \rightarrow C_2^* + M \qquad (3.39)$$

Carbon monoxide originated from HCOOH degradation and accumulated inside the bubble can also prevent H_2O_2 formation via reaction (3.31).

The sonochemistry of CO_2 was found to be very different from CO. Sonochemical activity of CO_2 has never been reported at low (20 kHz) ultrasonic frequency. At high frequency, addition of only 1% CO_2 to Ar causes a dramatic decrease in total sonoluminescence intensity of water [37, 38]. In terms of non-equilibrium plasma model of cavitation, this effect can be attributed to effective vibrational excitation of CO_2 molecules by collisions with electrons [35] which leads to rapid dissipation of electron energy. On the other hand, Henglein [39] and Harada [40] reported enhanced sonochemical dissociation of CO_2 at its very low content in argon (0.03–0.04 mol fraction). Carbon monoxide and formic acid were identified as principal products, but the yield of HCOOH was about 30 times smaller than the CO yield. Finally, the inhibiting effect on sonochemical oxidation of I^- ions in aqueous solutions increases with CO_2 concentration in CO_2/Ar gas mixture which most likely is related to the general decrease of sonochemical activity.

References

1. Richards WT, Loomis AL (1927) The chemical effects of high frequency sound waves. I. A preliminary survey. J Am Chem Soc 49:3086–3100
2. Schmitt FO, Johnson CH, O AR (1929) Oxidation promoted by ultrasonic radiation. J Am Chem Soc 51:370–375
3. Wu TY, Guo N, Teh CY, Hay JXW (2013) Advances in ultrasound technology for environmental remediation. Springer, Dordrecht
4. Gedanken A (2004) Using sonochemistry for the fabrication of nanomaterials. Ultrason Sonochem 11:47–55
5. Xu HX, Zeiger BW, Suslick KS (2013) Sonochemical synthesis of nanomaterials. Chem Soc Rev 42:2555–2567
6. Chave T, Navarro NM, Nitsche S, Nikitenko SI (2012) Mechanism of PtIV Sonochemical Reduction in formic acid media and pure water. Chem Eur J 18:3879–3885
7. Iida Y, Yasui K, Tuziuti T, Sivakumar M (2005) Sonochemistry and its dosimetry. Microchem J 80:159–164
8. Wood RJ, Lee J, Bussemaker MJ (2017) A parametric review of sonochemistry: control and augmentation of sonochemical activity in aqueous solutions. Ultrason Sonochem 38:351–370
9. Herzberg G (1979) Molecular spectra and molecular structure: constants of diatomic molecules. Van Nostrand, New York
10. Pflieger R, Brau HP, Nikitenko SI (2010) Sonoluminescence from $OH(C^2\Sigma^+)$ and $OH(A^2\Sigma^+)$ radicals in water: evidence for plasma formation during multibubble cavitation. Chem Eur J 16:11801–11803
11. Ebrahiminia A, Mokhtari-Dizaji M, Toliyat T (2013) Correlation between iodide dosimetry and terephthalic acid dosimetry to evaluate the reactive radical production due to the acoustic cavitation activity. Ultrason Sonochem 20:366–372
12. Ouerhani T, Pflieger R, Ben Messaoud W, Nikitenko SI (2015) Spectroscopy of sonoluminescence and sonochemistry in water saturated with N_2–Ar mixtures. J Phys Chem B 119:15885–15891
13. Abeledo CA, Kolthoff IM (1931) The reaction between nitrite and iodide and its application to the iodometric titration of these anions. J Am Chem Soc 53:2893–2897
14. Couto AB, de Souza DC, Sartori ER, Jacob P, Klockow D, Neves EA (2006) The catalytic cycle of oxidation of iodide ion in the oxygen/nitrous acid/nitric oxide system and its potential for analytical applications. Anal Lett 39:2763–2774
15. Mark G, Tauber A, Rudiger LA, Schuchmann HP, Schulz D, Mues A, von Sonntag C (1998) OH-radical formation by ultrasound in aqueous solution—Part II: Terephthalate and Fricke dosimetry and the influence of various conditions on the sonolytic yield. Ultrason Sonochem 5:41–52

16. Chang CY, Hsieh YH, Cheng KY, Hsieh LL, Cheng TC, Yao KS (2008) Effect of pH on Fenton process using estimation of hydroxyl radical with salicylic acid as trapping reagent. Water Sci Technol 58:873–879
17. Milne L, Stewart I, Bremner DH (2013) Comparison of hydroxyl radical formation in aqueous solutions at different ultrasound frequencies and powers using the salicylic acid dosimeter. Ultrason Sonochem 20:984–989
18. Nikitenko SI, Le Naour C, Moisy P (2007) Comparative study of sonochemical reactors with different geometry using thermal and chemical probes. Ultrason Sonochem 14:330–336
19. Pflieger R, Chave T, Vite G, Jouve L, Nikitenko SI (2015) Effect of operational conditions on sonoluminescence and kinetics of H_2O_2 formation during the sonolysis of water in the presence of Ar/O_2 gas mixture. Ultrason Sonochem 26:169–175
20. Mason TJ, Lorimer JP (1989) Sonochemistry, theory, applications and uses of ultrasound in chemistry. Prentice Hall, New Jersey
21. Petrier C, Jeunet A, Luche JL, Reverdy G (1992) Unexpected frequency-effects on the rate of oxidative processes induced by ultrasound. J Am Chem Soc 114:3148–3150
22. Beckett MA, Hua I (2001) Impact of ultrasonic frequency on aqueous sonoluminescence and sonochemistry. J Phys Chem A 105:3796–3802
23. Kanthale P, Ashokkumar M, Grieser F (2008) Sonoluminescence, sonochemistry (H_2O_2 yield) and bubble dynamics: frequency and power effects. Ultrason Sonochem 15:143–150
24. Ndiaye AA, Pflieger R, Siboulet B, Molina J, Dufreche JF, Nikitenko SI (2012) Nonequilibrium vibrational excitation of OH radicals generated during multibubble cavitation in water. J Phys Chem A 116:4860–4867
25. Navarro NM, Chave T, Pochon P, Bisel I, Nikitenko SI (2011) Effect of ultrasonic frequency on the mechanism of formic acid sonolysis. J Phys Chem B 115:2024–2029
26. Fischer CH, Hart EJ, Henglein A (1986) Ultrasonic irradiation of water in the presence of $^{18,18}O_2$—isotope exchange and isotopic distribution of H_2O_2. J Phys Chem 90:1954–1956
27. Petrier C, Combet E, Mason T (2007) Oxygen-induced concurrent ultrasonic degradation of volatile and non-volatile aromatic compounds. Ultrason Sonochem 14:117–121
28. Wagatsuma K, Hirokawa K (1995) Effect of oxygen addition to an argon glow-discharge plasma source in atomic-emission spectrometry. Anal Chim Acta 306:193–200
29. Shultes H, Gohr H (1936) Über chemische Wirkungen der Ultraschallwellen. Angew Chem 49:420–423
30. Misik V, Riesz P (1999) Detection of primary free radical species in aqueous sonochemistry by EPR spectroscopy. In: Crum LA, Mason TJ, Reisse JL, Suslick KS (eds) Sonochemistry and sonoluminescence, pp 225–236
31. Wakeford CA, Blackburn R, Lickiss PD (1999) Effect of ionic strength on the acoustic generation of nitrite, nitrate and hydrogen peroxide. Ultrason Sonochem 6:141–148
32. Hart EJ, Fischer CH, Henglein A (1986) Isotopic exchange in the sonolysis of aqueous-solutions containing $^{14,14}N_2$ and $^{15,15}N_2$. J Phys Chem 90:5989–5991
33. Pflieger R, Ouerhani T, Belmonte T, Nikitenko SI (2017) Use of NH ($A^3\Pi_i$-X^3 Σ^-) sonoluminescence for diagnostics of nonequilibrium plasma produced by multibubble cavitation. Phys Chem Chem Phys 19:26272–26279
34. Nikitenko SI, Martinez P, Chave T, Billy I (2009) Sonochemical disproportionation of carbon monoxide in water: evidence for treanor effect during multibubble cavitation. Angewandte Chemie-International Edition 48:9529–9532
35. Fridman A (2008) Plasma chemistry. Cambridge University Press
36. Bigeleisen J (1965) Chemistry of isotopes. Science 147:463−471
37. Navarro NM, Pflieger R, Nikitenko SI (2014) Multibubble sonoluminescence as a tool to study the mechanism of formic acid sonolysis. Ultrason Sonochem 21:1026–1029
38. Kumari S, Keswani M, Singh S, Beck M, Liebscher E, Deymier P, Raghavan S (2011) Control of sonoluminescence signal in deionized water using carbon dioxide. Microelectron Eng 88:3437–3441

39. Henglein A (1985) Sonolysis of Carbon-dioxide, nitrous-oxide and methane in aqueous-solution, Zeitschrift Fur Naturforschung Section B-a. J Chem Sci 40:100–107
40. Harada H (1998) Sonochemical reduction of carbon dioxide. Ultrason Sonochem 5:73–77

Conclusion

Quite a large knowledge has been gathered on cavitation bubbles in the past years, aiming at characterizing them from a physical and chemical point of view and at understanding the impact of various experimental parameters on their behavior and properties. Many works have dealt with bubble dynamics: measurements of the evolution of the bubble radius and of the bubble shape with time together with the development of more and more sophisticated models now allow to understand the concentration of energy at bubble collapse, the emission of acoustic waves, the formation of microjets, etc. Also the interactions between bubbles have been studied, interactions that lead to the formation of the observed bubble structures (clusters, filaments etc.). In parallel to the dynamics study of bubbles on the microscopic level, two other directions were pursued to characterize acoustic cavitation bubbles. The emission of light (sonoluminescence) at the bubble collapse, and chemical reactions (sonochemistry). Sonoluminescence gives an a priori direct insight into the conditions reached at collapse. Observed spectral features clearly indicate the formation of plasma that is not at equilibrium. Though more research is needed to properly characterize the formed plasma and understand its formation (in particular the electron temperature and density are mainly unknown), rovibronic temperatures of excited species were measured that indicate the large impact of, e.g., the ultrasonic frequency or the nature of the dissolved gas. These effects are in agreement with measurements of the sonochemical activity.

However, connections between the gained knowledge in each field remain poor, which is partly due to the different scales of study: Bubble observations (dynamics, shape, etc.) focus on the microscopic scale, while sonoluminescence and sono-chemistry deal with measurements on a macroscopic scale. Therefore, the latter two deliver an average picture of the effects of acoustic cavitation: averaged over space and also over time. A second difference is in the definition of the systems of interest: While measurements of bubble dynamics deal with a single or a few bubbles, sonochemical systems are composed of many bubbles. The bubble population is broad and covers bubbles of very different sizes, active or passive, spherical or strongly deformed, all of them interacting and evolving with time.

© The Author(s), under exclusive licence to Springer Nature Switzerland AG 2019   73
R. Pflieger et al., *Characterization of Cavitation Bubbles and Sonoluminescence*,
Ultrasound and Sonochemistry, https://doi.org/10.1007/978-3-030-11717-7

Furthermore, even the ultrasonic frequencies investigated can be different: High frequencies are often more interesting for chemists (due to the higher chemical activity provided), while measurements of bubble dynamics focus on low frequencies (favored due to bigger bubble sizes and longer acoustic periods).

Some technical limitations may be overcome in future years, which will help linking the knowledge gained from the various approaches (sonochemical activity, sonoluminescence spectroscopy, and bubble dynamics). First, observations of the bubbles may be extended toward higher frequencies if (even) faster cameras and special microscopic techniques are developed. Second, there is a need to go beyond the spatially and temporally averaged measurements of sonoluminescence spectra. Since the amount of emitted photons of the systems of interest for chemistry cannot be much increased, the sensitivity of the light detector should be improved to allow spatial resolution. The objective is to link spectral features, chemical activity, and the characteristics of each considered emitting bubble. As already observed (see Sect. 2.3.1) different types of bubbles can emit light, with different corresponding sonoluminescence spectra. The big challenge is now to get a full picture of which bubbles emit which spectra (that reflect the formed plasma in their core) and to link these observations to the chemical activity of each bubble. In a further step, one might think of a better control of the desired bubble populations and their dynamics.

In conclusion, cavitation and collapsing bubbles form a quite complex tool, employed since decades and not only for chemical applications. Nowadays, such systems are much better understood than in the early days, and the picture becomes more and more complete. Still there is plenty of work ahead, and we are convinced that the field will remain active and surprising also in the future.

Printed in the United States
By Bookmasters